Lecture Notes in Computer Science 4945

Commenced Publication in 1973
Founding and Former Series Editors:
Gerhard Goos, Juris Hartmanis, and Jan van Leeuwen

Stefan Lucks   Ahmad-Reza Sadeghi
Christopher Wolf (Eds.)

Research in Cryptology

Second Western European Workshop, WEWoRC 2007
Bochum, Germany, July 4-6, 2007
Revised Selected Papers

 Springer

Volume Editors

Stefan Lucks
Bauhaus-Universität Weimar, Fakultät Medien
Bauhausstr. 11, 99423 Weimar, Germany
E-mail: stefan.lucks@medien.uni-weimar.de

Ahmad-Reza Sadeghi
Ruhr-Universität Bochum, Institut für Systemsicherheit
Universitätsstr. 150, 44780 Bochum, Germany
E-mail: ahmad.sadeghi@trust.rub.de

Christopher Wolf
Ruhr-Universität Bochum
Horst-Görtz-Institut für Sicherheit in der Informationstechnik
Universitätsstr. 150, 44780 Bochum, Germany
E-mail: cbw@hgi.rub.de

Library of Congress Control Number: Applied for

CR Subject Classification (1998): E.3, D.4.6, I.1, K.6.5, K.4.4

LNCS Sublibrary: SL 4 – Security and Cryptology

ISSN   0302-9743
ISBN-10   3-540-88352-5 Springer Berlin Heidelberg New York
ISBN-13   978-3-540-88352-4 Springer Berlin Heidelberg New York

Springer is a part of Springer Science+Business Media

springer.com

© Springer-Verlag Berlin Heidelberg 2008
Printed in Germany

Typesetting: Camera-ready by author, data conversion by Scientific Publishing Services, Chennai, India
Printed on acid-free paper   SPIN: 12536580   06/3180   5 4 3 2 1 0

Preface

The Western European Workshop on Research in Cryptology (WEWoRC 2007) was the second of its kind. It was organized as a joint venture between the Horst Görtz Institute for Security in Information Systems (HGI), and the Special Interest Group on Cryptology (*FG Krypto*) of the German Computer Science Society (Gesellschaft für Informatik e.V.). The aim was to bring together researchers in the field of cryptology. The workshop focused on research from Masters and PhD students, and brought them together with more experienced senior researchers. The first workshop (WEWoRC 2005) was held in Leuven.

WEWoRC 2007 was held in the German Ruhr region, more particularly in Bochum, during July 4–6, 2007. Formerly a mining town, Bochum is currently growing into a knowledge-based economy. Aided by the city council, IT security is a special focus for economic development. Hence, it provided the perfect scenery for hosting this event. In total, we had 81 participants from 13 different countries (Belgium, Finland, France, Germany, Iran, Japan, Luxembourg, Malawi, Slovenia, Taiwan, Tunisia, UK, USA).

In total, we received 39 submissions of which 36 where chosen for presenting in 14 sessions. In addition, the program was enriched by two invited talks, namely, by George Danezis on *"Cryptography in Anonymous Communications"* and David Naccache on *"Products of Small Primes in Cryptography and Error-Correction."* Selecting papers for publication in these postproceedings was done in two phases. In the first phase, during the workshop, the authors of 24 of the 36 talks were invited to submit a full paper for these postproceedings. In the second phase, after we received the 24 invited submissions, these were reviewed by the members of our Program Committee. Each paper was reviewed in a careful refereeing process by at least three experts in the area. If one of the authors was a member of the Program Committee, at least five reviews were requested. We used a total of 73 reviews for finally selecting the 12 papers presented here.

We are very grateful to all the Program Committee members who devoted much effort and valuable time to reading and selecting the papers. These postproceedings contain the final versions of each paper revised after the conference. Since the revised versions were not checked by the Program Committee members rigorously, the authors must bear full responsibility for the contents of their papers. We also want to thank the external experts who assisted the Program Committee in evaluating various papers.

Special thanks to our sponsors who made it possible to offer WEWoRC for a competitive price. Their logos are on the first page of these post-proceedings. Similarly, we want to mention the cooperation with our academic partners EI-DMA and Ecrypt. In addition, we want to thank the local Organizing Committee for their skillful, professional, and enthusiastic support of WEWoRC. Keep in mind that all work was done voluntarily. Special thanks go in this context to the

Horst Görtz Institute, which kindly agreed to host the workshop in Bochum and for allowing us to use the HGI infrastructure (both technical and administrative) for WEWoRC.

Finally, we would like to thank all authors — including those whose submissions were not successful, as well as the workshop participants from around the world for their support, which made WEWoRC a big success.

December 2007   Stefan Lucks
                                                  Ahmad-Reza Sadeghi
                                                    Christopher Wolf

Organization

Program Committee

Ammar Alkassar	Sirix AG, Germany
Frederik Armknecht	NEC, Germany
N. Asokan	Nokia Research Helsinki, Finland
Roberto Avanzi	Ruhr University Bochum, Germany
Lynn Batten	Deakin University, Australia
Alex Biryukov	University of Luxembourg, Luxembourg
Johannes Blömer	Paderborn, Germany
Colin Boyd	Queensland University of Technology, Australia
Lejla Batina	KU Leuven, Belgium
Dario Catalano	CNRS-ENS, France; Universitá di Catania, Italy
Christophe Clavier	Gemalto, France
Jean-Sébastien Coron	University of Luxembourg, Luxembourg
Steven Galbraith	Royal Holloway, University of London, UK
Joachim von zur Gathen	b-it Bonn, Germany
Willi Geiselmann	TU Karlsruhe, Germany
Marc Girault	France Telecom, France
Louis Goubin	University of Versailles, France
Aline Gouget	Gemalto, France
Helena Handschuh	Spansion, France
Florian Hess	TU Berlin, Germany
Erwin Hess	Siemens, Germany
Ellen Jochemsz	TU Eindhoven, The Netherlands
Dogan Kesdogan	RWTH Aachen, Germany
Eike Kiltz	CWI, The Netherlands
Ulrich Kühn	Sirrix AG, Germany
Arjen Lenstra	EPFL, Switzerland
Francoise Levy-dit-Vehel	ENSTA, France
Gregor Leander	Ruhr University Bochum, Germany
Stefan Lucks	Bauhaus University Weimar, Germany
Keith Martin	Royal Holloway, University of London, UK
Alexander May	TU Darmstadt, Germany
Chris Mitchell	Royal Holloway, University of London, UK
David Naccache	ENS Paris, France
Heike Neumann	Philips Semiconductors, Germany
Svetla Nikova	KU Leuven, Belgium
Siddika Berna Ors	Istanbul Technical University, Turkey

Elisabeth Oswald   Bristol, UK
Christof Paar   Ruhr University Bochum, Germany
Kenny Paterson   Royal Holloway, University London, UK
Bart Preneel   KU Leuven, Belgium
Vincent Rijmen   TU Graz, Austria; Cryptomathic, Denmark
Ahmad-Reza Sadeghi   Ruhr University Bochum, Germany
Christian Tobias   Utimaco, Germany
Rei Safavi-Naini   University of Wollongong, Australia
Jörg Schwenk   Ruhr University Bochum, Germany
Nicolas Sendrier   INRIA, France
Stefaan Seys   KU Leuven, Belgium
Heiko Stamer   Kassel, Germany
Henk van Tilborg   TU Eindhoven, The Netherlands
Pim Tuyls   Philips, The Netherlands
Pascal Véron   University of Toulon, France
Moti Yung   Columbia University and RSA Laboratories,
                            USA
Michael Welschenbach   SRC Security Research and Consulting,
                            Germany
Ralf-P. Weinmann   TU Darmstadt, Germany
Thomas Wilke   Kiel, Germany
Ralph Wernsdorf   Rohde & Schwarz SIT GmbH, Germany
Christopher Wolf   PwC Luxembourg; K.U.Leuven, Belgium
Po-Wah Yau   Royal Holloway, University of London, UK
Erik Zenner   Technical University of Denmark

Referees

Ali Akbar Sobhi Afshar	Philippe Gaborit	Anja Korsten
Carlos Aguilar-Melchor	Sebastian Gajek	Paul Kubwalo
Mohammad Reza Aref	Steven Galbraith	Kerstin Lemke-Rust
Bechir Ayeb	Timo Gendrullis	Lijun Liao
Michael Beiter	Benedikt Gierlichs	Stéphane Manuel
Waldyr Benits	Tim Güneysu	Mark Manulis
Nicolas T. Courtois	Mabrouka Hagui	Gordon Meiser
Léonard Dallot	Rupert J. Hartung	David Naccache
George Danezis	Wei-Hua He	María Naya-Plasencia
Blandine Debraize	Marko Hölbl	Akira Otsuka
Jintai Ding	Hideki Imai	Christof Paar
Taraneh Eghlidos	Sebastiaan Indesteege	Souradyuti Paul
Mohammad Ehdaie	Kare Janussen	Selwyn Piramuthu
Thomas Eisenbarth	Fei-Ming Juan	Joris Plessers
Jan-Erik Ekberg	Timo Kasper	Bart Preneel
Junfeng Fan	Dalia Khader	Christoph Puttmann
Ewan Fleischmann	Hedi Khammari	Christian Rechberger

Andrea Röck
Ahmad-Reza Sadeghi
Kazuo Sakiyama
Sven Schäge
Dieter Schmidt
Thomas Schwarzpaul
Jörg Schwenk

Gautham Sekar
Nicolas Sendrier
Rie Shigetomi
Jamshid Shokrollahi
Dirk Stegemann
Yu-Ju Tu
Caroline Vanderheyden

Ingrid Verbauwhede
Tatjana Welzer
Fabian Werner
Ralph Wernsdorf
Rei Yoshida
Kazuki Yoshizoe

Silver Sponsors

Bronze Sponsors

Table of Contents

A Privacy Protection Scheme for a Scalable Control Method in Context-Dependent Services

Rei Yoshida[1], Rie Shigetomi[2], Kazuki Yoshizoe[1],
Akira Otsuka[2], and Hideki Imai[1,2]

[1] Chuo University, Imai Lab.,
Dept. of Electrical, Electronic, and Communication Engineering,
Faculty of Science and Engineering, Tokyo, Japan
{rei-yoshida,k-yoshizoe}@imailab.jp
[2] Research Center for Information Security,
National Institute of Advanced Industrial Science and Technology, Tokyo, Japan
{rie-shigetomi,a-otsuka,h-imai}@aist.go.jp
http://www.rcis.aist.go.jp/index-en.html

Abstract. Provision of context-dependent services is triggered when the context satisfies an execution condition. To deliver these services, users' contexts have to be determined by terminals such as GPS. However, GPS has efficiency (it must collect as many contexts as possible to provide services appropriately) and privacy problems (all data is concentrated on one place). Previous studies have addressed only either one of the two problems. We propose a scheme that protects users' privacy while maintaining efficiency by using the Randomized Response Technique.

Keywords: Context-dependent Service, Privacy Protection, Randomized Response Technique.

1 Introduction

Context is any information that can be used to characterize a person, a place, or an object's situation, or that is relevant to the interaction between a user and an application [3]. A context-dependent service changes according to the context of the user or his or her surroundings and supports his or her activity in the real world. The development of devices such as GPS, mobile phones, and non-contact IC cards makes these services possible. Examples are restaurant recommendation services, navigation systems for GPS-equipped mobile phones, and 24-hour home healthcare services.

This kind of service system consists of context acquisition terminals, servers, and users. Context acquisition terminals, such as mobile phones with GPS or sensor nodes, are connected with servers via a network. The server collects the values that the terminals acquired via the network and determines whether it is an appropriate time to execute the service by referring to the service's execution conditions. If the conditions are fulfilled, the service is provided to the user. Hereafter, *operation* means the flow from context acquisition to the provision of service.

S. Lucks, A.-R. Sadeghi, and C. Wolf (Eds.): WEWoRC 2007, LNCS 4945, pp. 1–12, 2008.

Context-dependent services have become popular because they are convenient and efficient. However, these services have two problems, efficiency and user privacy. The first problem is that the server has to monitor a huge number of contexts in real time to ensure appropriate service provision. This causes problems such as excessive computation load for the server and excessive bandwidth consumption for the network. The user's privacy is jeopardized when information about all the contexts, such as the user's name, location, and service provided, is collected to one place.

The simplest solution to the efficiency problem is the periodical communication method. However, if the frequency of the periodical method is low, it misses many appropriate chances to provide services. Uchida et al. proposed a system that overcomes the efficiency problem by using a Bayesian network, adjusting the communication cycle to eliminate unnecessary communication. We call this method BN and describe it in detail in section 4.2. The authors' simulation of the BN method shows that the method is ten times more efficient than the existing periodical method. However the BN method still had the privacy problem.

Therefore, we propose a system that solves both privacy and efficiency problems by using three controllers and Randomized Response Technique (RRT). This method uses a certification controller, an execution controller, and a state controller. Each controller controls one of three contexts: identity, service provided, or location history. Also, this simple segmentation method cannot update a Bayesian network's service provision information without violating privacy. However, the Bayesian network needs past information because next service provision is determined by this information. Our method uses RRT to collect information while protecting privacy. RRT can protect a user's privacy by randomizing each response according to a certain known probability distribution, and extract correct information because the randomized *lies* could be removed by the equation(1). Our proposed method can solve both the privacy and the efficiency problems of context-dependent services.

2   Examples

Examples of a context-dependent service are shown below:

1. **Restaurant recommendation service**

   These services automatically display restaurant recommendations on user terminals. The contexts they monitor are the user's distance from a restaurant and the number of available tables in the restaurant. The recommended restaurant must be within a certain distance and have an available table. If the user registers his preferences beforehand, the service may also be able to recommend a corresponding restaurant. If the recommended restaurant distributes coupons, the service can provide one to the user. Restaurants that correspond to the user's preferences are recommended based on these contexts. The problem of privacy arises due to the linkage of the user's location history, identity, and whether the user went to the recommended restaurant.

2. **Amusement park attraction recommendation service**
   This service automatically displays recommended attractions on the user's terminal. It monitors three contexts: the user's distance from attractions, wait times for attractions, and the authorization to enter certain attractions. The nearer the attraction, and the shorter the waiting time, the more highly recommended the attraction is. The service does not recommend attractions that the user cannot enter because of restrictions on the user's ticket. Attractions that the user can enter most easily are recommended based on these contexts. The problem of privacy arises from the linkage of the contexts for the user's location history, identity, preference for attractions and type of ticket.
3. **Management of a user's activity in an office**
   This service automatically locks or opens doors and limits computer access. The contexts it monitors are a user's location and authorized activity. If there are areas of a building that are off-limits to the user based on his level of authorization, the doors to those areas are open only when a user who has authorization for those areas is close by. If the user is away from his or her desk, the user's PC is locked to protect logins. The problem of privacy arises from the linkage of the user's location history, identity and level of authorization.

As mentioned above, context-dependent services face problems with high frequency context collection for real-time control and user privacy. Our method can solve these problems by adjusting the context collection cycle based on the Bayesian network and simple segmentation of each context: location, authority, and service execution.

3   Related Work

The problem of efficiency has been addressed in several previous studies [2,5,6,7]. Prabhakar et al. proposed a Q-index approach in which each mobile object is assigned a safe region [5]. Queries are only necessary when an object crosses the boundaries of a safe region. This method can reduce the number of unnecessary location updates. Hu et al. proposed an efficient algorithm for safe region computation [7]. Cai et al. proposed that the objects monitor the query region [6]. Part of a server's computation workload is distributed to the objects. However, these approaches can still violate a user's privacy because all contexts are gathered in one place.

The privacy problem has been addressed in a few previous studies [11,12,15]. Context Toolkit Project [11] is one of the first methods to address the privacy problem in context-aware systems. Context Toolkit consists of context widgets, interpreters, and aggregators. A special widget that acts as a gateway between other widgets and applications provides basic access control to protect privacy. Context Broker Architecture (CoBrA) [12] uses the Web ontology language, OWL, to support context reasoning. The central entity of this architecture is the server called the context broker. To protect user privacy, the broker enforces the privacy policies defined by users. However, these methods do not address efficiency. To provide context-dependent services in the real world, the problems of efficiency and privacy must be addressed.

There has been research on the problem of location privacy. The Mist protocol [13] solved the problem of location privacy by implementing a routing protocol combined with strong public key cryptography that allows the system to detect the presence of users in a place, but not to positively identify them. The Mix zone [14] introduces the concept of application zones and mix zones. When a user enters a mix zone, his or her identity is mixed with all others in the mix zone. Each time a user enters an application zone, he or she is assigned a new pseudonym. A third party application provider receives the pseudonym, and an untraceable user identity is associated with the user's location. Because these methods specialize in location privacy, it is difficult to use them for all context-aware services.

Roussaki et al. addressed both problems [15]. Their method trusts the service provider and aims to protect user's privacy from a third party. To protect privacy, the user's terminal encapsulates and segments his context information with pseudonyms. Also, the database servers are distributed based on location. Since most users request information related to their current location, the authors claimed the system is efficient. However, this method requires frequent *operations* on each local database server and its goal is not the same as ours.

4   Preliminaries

This section introduces the notation and building blocks of our scheme. In section 4.1, we notate some facts. In section 4.2, we introduce the BN method, which solves the efficiency problem. In section 4.3 we introduce the RRT privacy protection protocol.

4.1   Notation

Let k be a security parameter, q be a prime of length k, and $\mathbb{G}_q$ be a multiplicative cyclic group of order q. Assume that p is a large prime and q, $q|(p-1)$ is another prime. $\mathbb{Z}_q$ has a unique subgroup $\mathbb{G}$ of order q.

4.2   BN Method

This section explains the BN method, which solves the problem of efficiency using a Bayesian network.

Overview of BN method. In context-dependent services, constant location updates degrade efficiency. One solution to this problem is to execute control *operations* periodically. However, if the server executes *operations* at low frequency, service provision may be delayed. The BN method detects low risk period and reduce communications during such period. The risk detection is done by Bayesian network which is trained by the past data.

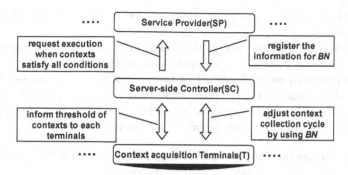

Fig. 1. Architecture of BN method

Figure 1 shows the architecture of the BN method. The BN method consists of service providers SP, the server-side controller, SC, and the context acquisition terminals, T. First, SP shows the service execution conditions to SC. Next, SC collects context values from T at appropriate times. These times are determined by calculating the probability of the execution conditions being satisfied and each context's effect on this probability. SC also informs each T of thresholds. T sends its value to SC when SC requests it and the value is higher than a threshold. Then, SC requests that SP provide the service when SC can verify that the execution conditions are satisfied. Finally, the requested SP provides the service to the user. Also, the Bayesian network is updated that the service has been provided.

The simulation of the BN method by Uchida et al. shows that the method is ten times more efficient than the existing periodical method.

Calculation of risk. The Bayesian network calculates the probability of service being provided based on how the contexts shift. The future context can be predicted by considering the current context with the Bayesian network.

For example, if only a small number of conditions are satisfied, there is a very small probability that all conditions will be satisfied in the next moment. In contrast, if most conditions are satisfied, there is significant risk of missing a chance to provide services. Therefore, their method reduces frequency of *operations* only when few conditions are satisfied.

Certain contexts have predictable values, others do not. For example, if it is known that a pedestrian user is in a certain place at a certain time and the user's speed and location are known, that user's location after one minute can be predicted with some certainty. Therefore, the predicted value of each context is used instead of its exact value to determine whether to provide a service. In contrast, a context such as the number of available tables at a restaurant is difficult to predict. Data used in the past are used to determine which contexts can be predicted with confidence.

However, a supposedly predictable context may shift unpredictably. For this reason, the server-side controller informs each context acquisition terminal of context thresholds. If a certain terminal detects that the context has reached a threshold, the terminal sends the context's value to the controller without the controller requesting it.

As mentioned above, the BN method eliminates unnecessary *operations* by configuring cycles for collecting contexts and setting thresholds for each terminal. The Bayesian network calculates the risk of missing chances to provide service based on data used in the past, improving accuracy of calculation by updating the Bayesian network with new information. For this reason, the Bayesian network has to be updated that services have been provided.

Definition. BN method consists of four algorithms that are BN_{setup}, $BN_{collect}$, $BN_{service}$, and *Learning*, which are defined below.

BN_{setup}: A probabilistic algorithm that, given information b to set up the Bayesian network, outputs the starting value of the Bayesian network Bn_0.

$$(Bn_0) \leftarrow BN_{setup}(b)$$

$BN_{collect}$: A probabilistic algorithm that, given the Bayesian network Bn, outputs C to collect.

$$(C) \leftarrow BN_{collect}(Bn)$$

$BN_{service}$: A probabilistic algorithm that, given the collected C and the Bayesian network Bn, immediately outputs an appropriate service S.

$$(S) \leftarrow BN_{service}(C, Bn)$$

Learning: A probabilistic algorithm that, given the existing Bayesian network Bn and the result of service execution R, outputs a new Bayesian network Bn'.

$$(Bn') \leftarrow Learning(Bn, R)$$

4.3   Randomized Response Technique

This section explains the Randomized Response Technique (RRT) protocol, which can collect statistical information while protecting user privacy.

Outline of RRT. RRT is commonly used to conduct polls on sensitive issues such as drug abuse and criminal records[8]. The underlying idea is for the respondent to randomize each response based on a certain known probability distribution. When evaluating all the answers to the poll, an interviewer can recover the true proportion of the poll by a simple method which is described in this section.

Respondents $\mathcal{R}$ answer a given question truthfully with some probability $P_{ct} > 1/2$ and *lie* with a fixed and known probability $1 - P_{ct}$. A *lie* is defined as the answer opposite to the answer of the *truth*. Assume that interviewers $\mathcal{I}$ cannot know whether an answer is the *truth* or a *lie*. Therefore, the $\mathcal{R}$'s privacy is protected because the $\mathcal{I}$ cannot connect the answer and the $\mathcal{R}$. Overall, if the P_{ct} is stabilized and $\mathcal{I}$ get a sufficient number of answers, $\mathcal{I}$ can statistically predict true answers with an easy calculation.

Definition. The general RRT flow is shown below. Participants in RRT are an interviewer $\mathcal{I}$ and respondents $\mathcal{R}$. This protocol consists of two algorithms: (RRT_{setup}, $RRT_{\mathcal{I}}$) and one interactive protocol: $RRT_{\mathcal{R}}$.

RRT_{setup}: A probabilistic algorithm that, given a security parameter k, outputs public key $pkey$ (Public key is not necessary in certain implementation of RRT) and P_{ct}. P_{ct} does not need to be set up here, but we do so for the sake of simplicity.

$$(pkey, P_{ct}) \leftarrow RRT_{setup}(1^k)$$

$RRT_{\mathcal{R}}$: An interactive protocol between $\mathcal{R}$ and $\mathcal{I}$ in which $\mathcal{R}$ and $\mathcal{I}$ are given public key $pkey$ and probability of true P_{ct}, and $\mathcal{R}$ is additionally given own answer t. In the protocol, $\mathcal{R}$ sends to $\mathcal{I}$ t *truth* with probability $P_{ct}1/2$ and encrypts t with $pkey$ if necessary. $\mathcal{I}$ checks probability P_{ct} if it is possible, decrypts t if needed, and collects information t'. $\mathcal{I}$ outputs t', which has probability P_{ct} of being a *lie*.

$$\langle t', \phi \rangle \leftarrow RRT_{\mathcal{R}} \langle \mathcal{I}(pkey, P_{ct}), \mathcal{R}(t, pkey, P_{ct}) \rangle$$

$RRT_{\mathcal{I}}$: Let π_A be the proportion of the population whose type is t to all answers N. A probabilistic algorithm that, given answers $t'_1, \ldots, t'_N$ and probability of true P_{ct}, outputs the true information π_A using the simple calculation below.

$$(\pi_A) \leftarrow RRT_{\mathcal{I}}(t'_1, \ldots, t'_N, P_{ct}, pkey)$$

Then, if we define

$$P_A = \pi_A P_{ct} + (1 - \pi_A)(1 - P_{ct}),$$

then π_A can easily calculate such that

$$\pi_A = \frac{P_A - (1 - P_{ct})}{2P_{ct} - 1}. \tag{1}$$

Cryptographic RRT. There are various methods for implementing RRT. This section explains one of them which uses Oblivious Transfer[9]. We decide to use this method because it has stability proof and can prevent some malicious acts of respondents.

RRT_{setup}: $\mathcal{I}$ generates random $a, b \leftarrow \mathbb{Z}_q$ and chooses $\sigma \in [1, n]$. $\mathcal{R}$ prepares n random bits $\mu_i \in 0, 1$ for $i \in [1, n]$, such that $\sum \mu_i = l$ if $t = 1$ and $\sum \mu_i = n - l$ if $t = 0$ when $P_{ct} = l/n$. Additionally, $\mathcal{R}$ sets $\mu_{n+1} = 1 - t$.

$RRT_{\mathcal{R}}$: $\mathcal{I}$ sends (A,B,C)$(g^a, g^b, g^{ab-\sigma+1})$ to $\mathcal{R}$. For $i \in [1, n]$ $\mathcal{R}$ generates random (r_i, s_i), computes $w_i g^{r_i} A^{s_i}$, computes an encryption y_i of μ_i using $v_i B^{r_i} E(Cg^{i-1})^{s_i}$ as a key, and sends (w_i, y_i) to $\mathcal{I}$.

$RRT_{\mathcal{I}}$: $\mathcal{I}$ computes $w_\sigma^b = v_\sigma$ and decrypts y_σ using v_σ as the key, obtaining μ_σ. $\mathcal{I}$ verifies the value of l and halts if the verification fails.

Using this method, $\mathcal{R}$ cannot know which answer $\mathcal{I}$ is going to use, and $\mathcal{I}$ cannot know $\mathcal{R}$'s real answer. In addition, only when $\mathcal{R}$ makes a correct input that includes assigned proportion of *lies*, $\mathcal{R}$ can receive services. If $\mathcal{R}$ makes an input with illegal proportion of *lies* (such as a 100% *lie*), $\mathcal{I}$ can determine that it is wrong.

5 Proposed System

Here, we explain our proposed method for solving the privacy and efficiency problems.

5.1 Architecture of Proposed Method

We begin by explaining our system. Figure 2 shows the architecture of the method.

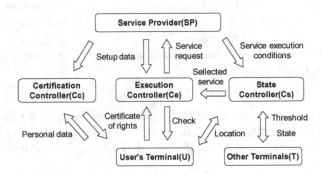

Fig. 2. Architecture of proposed method

The entities are service providers $\mathcal{SP}$, context acquisition terminals $\mathcal{T}$, user's terminals $\mathcal{U}$, the certification controller $\mathcal{C}_c$, the execution controller $\mathcal{C}_e$, and the state controller $\mathcal{C}_s$. $\mathcal{C}_c$ is entrusted to the trusted third certification authority. $\mathcal{C}_s$ is the server that comprehends whole states, such as the BN method's controller. $\mathcal{C}_e$ is a controller for verification of certification and determination of service execution, such as applications provided from $\mathcal{SP}$, to each user. To safeguard privacy, our method separates user identity, location history, and service history. To protect privacy while maintaining efficiency, we use a simple segmentation in which $\mathcal{C}_c$, $\mathcal{C}_s$, and $\mathcal{C}_e$ each have responsibility for one of these segments. This simple segmentation is natural because the certification controller is a trusted certification authority such as a credit company, the execution controller is a user's application, and the state controller is a server such as a mobile phone company. Therefore, we assume that these three entities are separated.

5.2 Requirements

Our proposed method fulfills the following requirements.

- **Correctness:**
  The service provider can provide an appropriate service to the user only if the user has the legitimate right to receive the service.

- **Efficiency:**
  The service provider can execute services at appropriate times based on real-time monitoring when redundant *operations* have been eliminated based on predictions.
- **Privacy:**
  Anonymity: Only the user must be able to connect his or her own personal data with the records of services provided to him or her. SP can only know whether the user has the right to receive the service.
  Nontraceability: Only the user can connect his or her personal data with his or her location history. C_s must be unable to connect location history with the user's rights and service history.
  History separation: It must not be able to connect service execution history with the user's location history. However, this requirement is difficult to achieve, because these histories must be connected when training the Bayesian network.

5.3 Construction

The flow of our system is shown below. The participants in our systems are the service provider SP, context acquisition terminals T and the user's terminal U, the certification controller C_c, the execution controller C_e, and the state controller C_s. Our method consists of two algorithms ($Certification_{issue}$, $Certification_{verify}$) in addition to the BN method and RRT protocol.

Initialization. Set up security parameter k, $pkey$, which is a public key, and probability P_{ct} using RRT. $(pkcy, P_{ct}) \leftarrow RRT_{setup}(1^k)$.

Setup. SP shows which users each service can be provided to C_c. SP shows to C_e, P_{ct} and the location each service can be executed. SP shows the information, b, needed to set up the Bayesian network to C_s and generates Bn_0, which is initial value of the Bayesian network, Bn. $(Bn_0) \leftarrow BN_{setup}(b)$.

Issuance of right. C_c issues certificates which show the user's rights to receive services, to U as a context of user's authority (CC). $(CC) \leftarrow Certification_{issue}(U)$.

Service recommendation. C_s collects C', which is context without personal data, at appropriate times $(C') \leftarrow BN_{collect}(Bn)$ and informs each terminals of acceptable values of contexts as thresholds in the same way as BN method's controller. T sends the value to C_s when the value is higher than a threshold value. When C_s has checked whether the execution conditions are satisfied, C_s informs C_e that service S satisfied all execution conditions. $(S) \leftarrow BN_{service}(C', Bn)$.

Service execution. C_e gets CC issued by C_c as context of the U's authority. When U has the authority to be provided the S, C_e requests the service execution of SP. $(t) \leftarrow Certification_{verify}(S, CC)$. SP provide the service to U.

Learning. C_e sends t of determination of the service execution to C_s using RRT and C_s receives t'. $\langle t', \phi \rangle \leftarrow RRT_R \langle C_s(pkey), C_e(t, pkey, P_{ct}) \rangle$. C_s gets true information π_A by calculation $(\pi_A) \leftarrow RRT_{\mathcal{I}}\ (t'_1, \ldots, t'_N, P_{ct}, pkey)$ and updates Bn' with it. $(Bn') \leftarrow Learning(Bn, \pi_A)$.

However, this update is executed at an appropriate time (when there will be small effect on service execution).

6   Discussion

6.1   Requirements

Here, we discuss the requirements of **Correctness**, **Efficiency**, and **Privacy**.

Correctness. The user's rights to receive services are checked by a certificate, which is issued by a trusted third party. Because the certificate is assumed to be trustworthy, **Correctness** is achieved.

Efficiency. Our method only checks on the user's right to receive services upon each service execution. Also, the certification controller pre-certifies authority, so it has little effect on efficiency. It only increases the number of communications between the state controller and the execution controller. Therefore, **Efficiency** depends on the BN method.

Our method uses RRT for sending information from the execution controller to the state controller and while protecting privacy. True information can be collected from the set of information that includes *lies*, and **Efficiency** improves as the amount of collected information increases.

In this instance, the amount of calculation our method must do increases due to use of CRRT using Oblivious Transfer. However, if the information is collected at midnight, when there is small effect on service, the **Efficiency** surpasses that of the old simple frequency method while maintaining reasonable service quality. The other problem of information gathering is discussed in section 6.2.

Privacy. Here, we discuss whether our method satisfies the **Privacy** requirement. As discussed in section 5.1, our method separates information.

Anonymity: The server cannot identify a user to whom the server is providing service. The certification controller cannot access information about service execution. This maintains user anonymity.

Nontraceability: The state controller cannot identify a user from whom the state controller collects contexts. The certification controller cannot access information about a user's location history. Therefore, the user is untraceable.

Learning-independence: The state controller is not able to access information about service execution without RRT. It is assumed that RRT can separate service provision and location history. Also, the server cannot access a user's location history from the state controller. Because as location history and execution history are periodically processed, information could not be extracted using timing. Therefore, each kind of record is independent of the others.

6.2 History Separation

The probability that service will be provided is predicted and set up based on data used in the past by a Bayesian network, which needs to be updated whenever service is provided.

The execution controller informs the state controller of information it has collected with RRT. The state controller cannot know which information is true; it can only check the percentage of *lies*. Also, the state controller can get true information by using RRT. Therefore, if there are a sufficient number of samples, the state controller can update the Bayesian network. However, the state controller must monitor all users without revealing personal data, and the particular service cannot be reflected to the Bayesian network.

In this paper, we assume the use of CRRT with Oblivious Transfer. However any implementation of RRT is affordable if it has the ability to prevent the user from manipulating the probability of *lies*.

7 Conclusion

We proposed a method that solves both the efficiency and privacy protection problems of context-dependent services. For efficiency, we rely on the state controller's efficiency. We used the BN method, which is described in [2], because it is the most efficient method currently known. To solve the privacy problem, we use three controllers to separate the user's identity, location history, and service history. This simple segmentation does not affect efficiency because these functions are normally separated, but there is still the problem of updating the Bayesian network with service provision information. We propose using RRT between C_s and C_e, which allows BN to be updated while protecting privacy and maintaining efficiency.

References

1. Yoshida, R., Shigetomi, R., Imai, H.: A Privacy Protection Scheme for a Scalable Control Method in Context-dependent Services. In: Proc. Society of Information Theory and Its Applications, SITA 2006, pp. 327–330 (2006) (in Japanese)
2. Uchida, W., Kasai, H., Kurakake, S.: A Scalable Execution Control Method for Context-dependent Services. In: Proc. IEEE Int'l Conf. on Pervasive Services (ICPS 2006), pp. 121–130 (2006)
3. Dey, A.K., Abowd, G.D., Salber, D.: A Conceptual Framework and a Toolkit for Supporting the Rapid Prototyping of Context-Aware Applications. HCI Journal 16(2-4), 97–166 (2001)
4. Baldauf, M., Dustdar, S., Rosenberg, F.: A survey on context-aware systems. Int. J. Ad Hoc and Ubiquitous Computing 2(4), 263–277 (2007)
5. Prabhakar, S., Xia, Y., Kalashnikov, D., Aref, W.G., Hambrusch, S.: Query indexing and velocity constrained indexing:Scalable techniques for continuous queries on moving objects. IEEE Transactions on Computers 15(10), 1124–1140 (2002)

6. Cai, Y., Hua, K.A., Cao, G.: Processing Range-Monitoring Queries on Heterogeneous Mobile Objects. In: Proc. 2004 IEEE Int'l Conf. on Mobile Data Management(MDM 2004), pp. 27–38 (2004)
7. Hu, H., Xu, J., Lee, D.L.: A Generic Framework for Monitoring Continuous Spatial Queries over Moving Objects. In: Proc. 2005 ACM SIGMOD Int'l Conf. on Management of Data, pp. 479–490 (2005)
8. Warner, S.L.: Randomized Response:A Survey Technique for Eliminating Evasive Answer Bias. Journal of the American Statistical Association 60(309), 63–69 (1965)
9. Ambainis, A., Jakobsson, M., Lipmaa, H.: Cryptographic Randomized Response Technique. In: Bao, F., Deng, R., Zhou, J. (eds.) PKC 2004. LNCS, vol. 2947, pp. 425–438. Springer, Heidelberg (2004)
10. Kikuchi, H., Akiyama, J., Nakamura, G., Gobioff, H.: Stochastic Voting Protocol To Protect Voters Privacy. In: 1999 IEEE Workshop on Internet Applications, July 26-27, pp. 103–111 (1999)
11. Salber, D., Dey, A.K., Abowd, G.D.: The Context Toolkit: Aiding the Development of Context-Enabled Applications. In: Conference on Human Factors in Computing Systems 1999 (May 1999)
12. Chen, H., Finin, T., Joshi, A., Kagal, L.: Intelligent Agents Meet the Semantics Web in Smart Spaces. IEEE Internet Computing 8(6), 69–79 (2004)
13. Al-Muhtabi, J., Cambell, R., Kapadia, A., Mickunas, D., Yi, S.: Routing through the mist:Privacy preseving communication in ubiquitous computing environments. In: Int'l conf of Distributed Computing Systems(ICDCS 2002), Vienna, Austria (July 2002)
14. Beresford, A., Stajano, F.: Mix zones:User privacy in location-aware services. In: IEEE Annual Conf.e on pervasive Computing and communications Workshops, Florida, USA (March 2004)
15. Roussaki, I., Strimpakou, M., Pils, C., Kalatzis, N., Neubauer, M., Hauser, C., Anagnostou, M.: Privacy-Aware Modelling and Distribution of Context Information in Pervasive Service Provision. In: Proc. IEEE Int'l Conf. on Pervasive Services (ICPS 2006), pp. 150–160 (2006)
16. IST Daidalos Research, http://www.ist-daidalos.org/

The GPS Identification Scheme Using Frobenius Expansions*

Waldyr D. Benits Junior** and Steven D. Galbraith***

Mathematics Department,
Royal Holloway University of London,
Egham, Surrey TW20 0EX, UK
{w.benits-junior,steven.galbraith}@rhul.ac.uk

Abstract. The Girault-Poupard-Stern (GPS) identification scheme is designed for public key cryptography on very restricted devices. We propose a variant of GPS for Koblitz elliptic curves using Frobenius expansions. The idea is to use Frobenius expansions throughout the protocol, so there is no need to convert between integers and Frobenius expansions. We give a security analysis of the proposed scheme.

Keywords: Elliptic Curves, Frobenius expansions, GPS identification scheme.

1 Introduction

The GPS public key identification scheme is a three move challenge-response protocol based on the Schnorr signature scheme [16]. It was first described by Girault [9] and later developed by Poupard and Stern [14] (also see [8]).

Recall that a public key identification scheme allows a prover to convince a verifier that she possess the private key. In a three move protocol, the prover sends a commitment (which can be computed in advance, i.e. offline), then receives a challenge and answers with a response (this is the "online step", which must be performed in realtime). The verifier then performs a computation involving the commitment, challenge, response and public key and outputs either *accept* or *reject*.

The idea of the GPS scheme is to make the online phase as fast and simple as possible, so that it can be easily performed by very low power devices. Further improvements to speed-up the online phase were proposed by Girault and Lefranc [10]. Okamoto, Katsuno and Okamoto [13] give an approach to reduce the bandwidth of the online step at the expense of the size of the public key.

A typical application of GPS (see [8]) is "on-the-fly" authentication at a road toll. Each car has a low cost smart card in which a GPS protocol runs. The

* The work described in this paper has been supported in part by the European Commission through the IST Programme under Contract IST-2002-507932 ECRYPT.
** This author thanks the Brazilian Navy for support.
*** This author supported in part by EPSRC Research Grant EP/D069904/1.

S. Lucks, A.-R. Sadeghi, and C. Wolf (Eds.): WEWoRC 2007, LNCS 4945, pp. 13–27, 2008.
© Springer-Verlag Berlin Heidelberg 2008

time required to transmit data and to perform online calculations is very small, hence the car does not need to stop at the toll to be authenticated. The offline operations may be calculated while the car is driving along. Note that we assume that the verifier (i.e., toll gate) has significant computational resources, so that the verification step can be done quickly.

Many of the original proposals for the GPS scheme were based in $(\mathbb{Z}/N\mathbb{Z})^*$ where N is an RSA modulus, or $\mathbb{F}_p^*$ where p is a large prime. However, since the computational device has extremely limited power it is more natural to work with elliptic curves, especially Koblitz curves over finite fields of small characteristic. The GPS protocol as described in [8] can be implemented with such elliptic curves, and the security results apply to this case. Using elliptic curves can give a significant speedup to the offline generation of the commitment and the verification step, as well as having lower memory and power consumption requirements. Due to the nature of the GPS protocol, using elliptic curves does not have any effect on the running time of the online step.

One can speed up the offline operations on Koblitz curves significantly by using Frobenius expansions to compute the required point multiplications. However, to implement the GPS protocol would require conversion between Frobenius expansions and integers and this would lead to extra code on the device (i.e., silicon area) and extra computational cost. Hence, the motivation of the present paper is to develop a GPS system which uses Frobenius expansions throughout. This will lead to fast and simple offline operations while still keeping the online operation fast (though the arithmetic of the online operation is more complicated than standard integer arithmetic and so is not as fast as standard GPS).

2   The Original GPS Scheme

The original Schnorr scheme [16] uses an element $g \in \mathbb{Z}_p^*$ of prime order q dividing $p-1$. A prover holds a private key $s \in \mathbb{Z}_q$ and a public key $I = g^{-s} \pmod{p}$. In the three step protocol, the prover generates a secret random r, computes the **commitment** $X = g^r \pmod{p}$ and sends X to the verifier. The verifier then generates a random **challenge** c and sends it to the prover. After receiving c, the prover computes the **response** $y = r + sc \pmod{q}$ and sends y to the verifier. The computation of y is the only online step. Finally, the verifier checks whether or not $X \equiv g^y I^c \pmod{p}$. If $X \not\equiv g^y I^c \pmod{p}$, the verifier rejects the proof. This round can be repeated l times, but in practice, we usually take $l = 1$.

The main idea of the GPS protocol is to eliminate the modular reduction performed during the response step. In other words, the response is just $y = r + sc$. This makes the computation more efficient, and it reduces the code footprint on the device (since there is no need to implement arithmetic modulo q). The modular reduction in the computation of y is used to prove the zero knowledge property of Schnorr signatures, hence a new proof of security is required for GPS and the parameters have to be carefully chosen (see [8, 14]). In [8] the interactive protocol is proven to have statistical zero knowledge. The original proposal by

Girault [9] used the group $(\mathbb{Z}/N\mathbb{Z})^*$ where N is an RSA modulus. Later work [8] proposed any cyclic group for which the discrete logarithm problem is hard.

An improvement to the original scheme, for which the online step is just a single addition, was presented by Girault and Lefranc [10]. It requires that the challenge (or alternatively, the private key) have a specific sparse form. The below table gives the reader an idea of the bitlengths of the integers (s, c, r) for an 80-bit security level of the private key and probability of successful forgery at most $1/2^{35}$. For more details see [8, 10].

Table 1. Numerical Example of GPS scheme

Scheme	bitlength of s	bitlength of c	bitlength of y
Standard GPS	160	35	275
Girault-Lefranc	160	940	1180

3   Koblitz Curves and Frobenius Expansions

In this section, we briefly review some properties of Koblitz elliptic curves [11] and how we can perform point multiplication efficiently using Frobenius expansions. A *Koblitz curve* is an elliptic curve E defined over a small finite field $\mathbb{F}_q$ such that the group $E(\mathbb{F}_{q^m})$ is suitable for cryptography for some $m > 1$. The most popular choice is curves over $\mathbb{F}_2$ and so we give the details in this case only.

Let $E_a : y^2 + xy = x^3 + ax^2 + 1$, with $a \in \{0, 1\}$ be an elliptic curve defined over $\mathbb{F}_2$. Denote by $\mathcal{O}$ the point at infinity. Let $E_a(\mathbb{F}_{2^m})$ be the group of $\mathbb{F}_{2^m}$-rational points on E_a. We assume that $\#E_a(\mathbb{F}_{2^m})$ has a large prime factor. The 2-power Frobenius map is

$$\tau : E_a(\mathbb{F}_{2^m}) \to E_a(\mathbb{F}_{2^m})$$
$$(x, y) \mapsto (x^2, y^2)$$
$$\mathcal{O} \mapsto \mathcal{O}.$$

Let $t = (-1)^{1-a}$. Then τ satisfies the equation $\tau^2(P) - [t]\tau(P) + [2]P = \mathcal{O}$ for all points $P \in E(\mathbb{F}_{2^m})$ (see [11]). In other words,

$$[2]P = [t]\tau(P) - \tau(\tau(P)).$$

Note also that τ satisfies $\tau^m(P) = P$ for all $P \in E(\mathbb{F}_{2^m})$.

Instead of computing $[n]P$ for a large integer n, one can represent n as a *Frobenius expansion*

$$\sum_{i=0}^{\mathcal{N}} n_i \tau^i, \qquad n_i \in \{-1, 0, 1\}.$$

See Solinas [20] for an algorithm to write n in this form. One can then compute $[n]P$ as $\sum_{i=0}^{\mathcal{N}} n_i \tau^i(P)$. In practice, the computation can be twice as fast as the standard double-and-add scalar multiplication (see [12] for the comparison).

We call two Frobenius expansions a, b *equivalent* if $[a]P = [b]P$ for all $P \in E(\mathbb{F}_{2^m})$ and write $a \equiv b$. Note that Frobenius expansions are not unique (for example, $1 \equiv -1 + t\tau - \tau^2$ and $1 \equiv \tau^m$) though one can use a non-adjacent form of degree $< m$ which is unique (see Solinas [19, 20]).

Definition 1. *For $n \in \mathbb{N}$ we define the set of τ-adic expansions of length n to be*

$$\mathcal{T}_n = \left\{ x_0 + x_1\tau + x_2\tau^2 + \cdots + x_{n-1}\tau^{n-1} \mid x_j \in \{-1, 0, 1\} \right\}. \tag{1}$$

For $x \in \mathcal{T}_n$ let $i \leqslant n - 1$ be the largest index such that $x_i \neq 0$. We say that x has degree i, denoted $\deg(x) = i$. The number of non-zero coefficients x_j is called the weight.

We emphasize that we consider elements of $\mathcal{T}_n$ as polynomials and not as endomorphisms. We interchangeably use the terminology *Frobenius expansion* and *τ-adic expansion* for such polynomials.

One can store a τ-adic expansion as a bitstring using Huffman encoding: for example coefficient 0 is represented as 0, coefficient 1 as 10 and coefficient -1 as 11. Hence a τ-adic expansion of length n and weight w requires $n + w$ bits (hence, we typically require $\approx 5n/3$ bits for a random τ-adic). If a non-adjacent form[1] is used then this can be shortened to $n + 1$ bits.

We now briefly mention the algorithms to convert between Frobenius expansions and integers. Given an integer n, Algorithm 1 of [20] outputs a Frobenius expansion in non-adjacent form equivalent to n. This algorithm only involves elementary integer operations. One problem is that the resulting Frobenius expansion typically has degree much larger than the original bitlength of n, so Solinas gives a method (Routine 62 in Section 5.2 of [20]) to output the remainder after division by $\tau^m - 1$. This latter algorithm requires many integer divisions. To convert a Frobenius expansion $\sum_i n_i\tau^i$ to an integer: let q be the order of P, compute the eigenvalue of Frobenius λ on $\langle P \rangle$ (λ is a root of $x^2 - t_m x + 2^m$ (mod q) where t_m is the trace of the 2^m-power Frobenius) and then compute $n = \sum_i n_i\lambda^i$ modulo q. In other words, this algorithm requires arithmetic modulo a large prime.

3.1   Arithmetic on τ-Adic Expansions

For our protocol we will need to add and multiply Frobenius expansions. The subtlety is that we require the result to have coefficients only in $\{-1, 0, 1\}$. Any coefficients which are outside this set must be reduced using the relation $2 = t\tau - \tau^2$, and this can lead to an increase in the degree of the expansion. There are two options for performing arithmetic:

1. Compute with coefficients in $\mathbb{Z}$ and reduce to coefficients in $\{-1, 0, 1\}$ at the end;
2. Perform the basic arithmetic operations so that all values have coefficients in $\{-1, 0, 1\}$ at all times.

---

[1] A non-adjacent form is when no two consecutive coefficients can both be non-zero.

Algorithm 1 presents an efficient addition algorithm of the second type. The important feature of this algorithm is that, although the carry values c_0, c_1 are integers, the arrays a, b and d only ever contain entries in the set $\{-1, 0, 1\}$. Hence this algorithm may be suitable for implementation on smart cards.

Algorithm 1. Efficient τ-adic addition

System Parameters: Frobenius polynomial $\tau^2 - t\tau + 2$ where $t = \pm 1$.
Input: τ-adic expansions a and b.
Output: $d \equiv (a + b)$.
1. $\deg = \max(\deg(a), \deg(b))$
2. $c_0 = 0$, $c_1 = 0$, $i = 0$ and $d = 0$
3. **while** $(i \leqslant \deg)$ or $(c_0 \neq 0)$ or $(c_1 \neq 0)$ **do**
4. $x = a[i] + b[i] + c_0$
5. $c_0 = c_1$ and $c_1 = 0$
6.   **if** $(x < -1)$ or $(x > 1)$ **then**
7. $q = \text{sign}(x) \times (|x| \text{ div } 2)$
8. $x = x - 2q$
9. $c_0 = c_0 + tq$
10. $c_1 = c_1 - q$
11.   **end if**
12.   [Optional randomisation step]
13. $d[i] = x$
14. $i = i + 1$
15. **end while**
16. **Return** d

The optional randomisation step in line 12 will be needed in the protocol. It is the following: if i is less than some parameter $\mathcal{K}'$ then generate a uniform random $b \in \{-1, 0, 1\}$ and arrange that $x = b$ by adding $(b - x)(\tau^i - \tau^{m+i})$. This corresponds to taking an equivalent representation for the sum. Note that this operation may increase the degree of the result by m and requires storing a 'carry' to be included when computing the $(m + i)$-th term[2]. Note that the randomisation method of Ebeid and Hasan [6] does not give uniform output of the low-degree coefficients and so it is not sufficient for our protocol.

Even with the non-randomised version, one sees that the degree can increase significantly if the carry values c_0, c_1 take non-zero values of large enough absolute value. Indeed, from Algorithm 1 one can derive a non-deterministic method to construct, given a τ-adic a, a τ-adic b such that $\deg(a + b) > \deg(b) \geqslant \deg(a)$ (note that the difference $\deg(a+b) - \deg(b)$ is bounded for fixed a). For example, given $a = 1 + \tau^3 - \tau^5$ one can choose $b = -\tau^5 + \tau^7$ so that the result of computing $a + b$ using Algorithm 1 is $1 + \tau^3 - t\tau^6 + t\tau^8 - \tau^9$. Indeed, it is clear that one can choose the first $\deg(a)$ coefficients of b at random, so there are at least $3^{\deg(a)}$ such choices for b (due to the choices available when constructing b there are

---

[2] We are using the polynomial $1 - \tau^m$, but could use any polynomial equivalent to zero and with constant coefficient 1.

many more choices for b, see heuristic 1 below). Similarly, one can choose b such that $\deg(a+b) < \deg(a)$ (for example, for the above a take $b = -\tau^3 + \tau^5$ so that $a + b = 1$); again there are about $3^{\deg(a)}$ possible choices.

Our variant of the GPS protocol will require computing $y = r + sc$ where r, c and s are all Frobenius expansions. The multiplication $s \times c$ can be performed by repeated shifting and addition using Algorithm 1 (possibly combined with a Karatsuba approach). Note that the arithmetic coming from Algorithm 1 is not associative (for example $(1+1) + -1 \neq 1 + (1 + -1)$) though all results are equivalent. There are several natural algorithms for computing sc and, due to lack of associativity, they will generally give different results. For the analysis of the protocol we assume that a fixed algorithm is used for computing sc in the protocol and the proof of Theorem 2.

Clearly, the product sc can have degree larger than $\deg(s) + \deg(c)$ but in practice it is not very much bigger. We have performed a number of experiments using MAGMA [5]: For $(\mathcal{C}, \mathcal{S}) = (23, 160)$ and randomly chosen $c \in \mathcal{T}_{\mathcal{C}}$ and $s \in \mathcal{T}_{\mathcal{S}}$ we found that $\deg(s \times c) \leqslant \deg(s) + \deg(c) + 5$ with high probability (for more details of our experimental results see [2]).

Definition 2. *For $\mathcal{S}, \mathcal{C} \in \mathbb{N}$, we define*

$$\Phi = (\mathcal{S} - 1) + (\mathcal{C} - 1) + 5 = \mathcal{S} + \mathcal{C} + 3$$

and

$$\mathcal{T}_{\Phi, \mathcal{R}} = \left\{ y \in \mathcal{T}_{\mathcal{R}} : y_n \neq 0 \text{ for some } n \text{ such that } \Phi \leqslant n < \mathcal{R} \right\}.$$

Our protocol will compute $r + sc$ where $\deg(r)$ is much bigger than $\deg(s) + \deg(c)$ and it is important to ensure that the degree is not likely to increase. For a given choice of sc of degree Φ there seem to be $\leqslant 3^{\Phi} \times 2^{\mathcal{K}-1}$ values for r satisfying $\deg(r) < \Phi + \mathcal{K}$ and $\deg(r+sc) \geqslant \Phi + \mathcal{K}$. Hence, the probability that $\deg(r+sc) \geqslant \Phi + \mathcal{K}$ can be estimated as $\frac{1}{3}(2/3)^{\mathcal{K}-1} = 1/3^{1+\log_3(3/2)(\mathcal{K}-1)} \approx 1/3^{0.63+0.37\mathcal{K}}$. If $\mathcal{K}$ is sufficiently large then the probability of this event for randomly chosen r is negligible. Our experiments showed that for $(\mathcal{C}, \mathcal{S}, \mathcal{K}) = (23, 160, 20)$, the probability of $\deg(r + sc) \neq \deg(r)$ over random $c \in \mathcal{T}_{\mathcal{C}}$, $s \in \mathcal{T}_{\mathcal{S}}$ and $r \in \mathcal{T}_{\Phi + \mathcal{K}}$ is approximately $0.00004 \leqslant 1/3^{0.37 \times 20 + 0.63} \approx 0.0001$.

Hence we propose the following heuristic. We say that a probability P is *negligible* if it decays exponentially (e.g., $O(1/3^{\mathcal{K}})$) and is *overwhelming* if $1 - P$ is negligible.

Heuristic 1. *Fix $s \in \mathcal{T}_{\mathcal{S}}$. Suppose $\mathcal{K} \in \mathbb{N}$ is sufficiently large and let $\mathcal{R} = \Phi + \mathcal{K}$. Then the probability over $r \in \mathcal{T}_{\mathcal{R}}$ that there exists some $c \in \mathcal{T}_{\mathcal{C}}$ such that $\deg(r + sc) \geqslant \mathcal{R}$ is negligible. More precisely,*

$$\#\{r \in \mathcal{T}_{\mathcal{R}} : \exists\, c \in \mathcal{T}_{\mathcal{C}} \text{ with } \deg(r + sc) \geqslant \mathcal{R}\} = O(3^{\Phi} 2^{\mathcal{K}-1}) = \tilde{O}(3^{\Phi + 0.63\mathcal{K} - 0.63}).$$

We also need to know how likely it is that $\deg(r+sc) \leqslant \Phi$. The above arguments indicate there should be at most 3^{Φ} such choices for r. Thus, we can state the following heuristic:

Heuristic 2. *Fix $s \in \mathcal{T}_S$. Suppose $\mathcal{K} \in \mathbb{N}$ is sufficiently large and let $\mathcal{R} = \Phi + \mathcal{K}$ Then the probability over $r \in \mathcal{T}_\mathcal{R}$ that there exists some $c \in \mathcal{T}_C$ such that $\deg(r + sc) < \Phi$ is negligible. More precisely,*

$$\#\{r \in \mathcal{T}_\mathcal{R} : \exists\, c \in \mathcal{T}_C \text{ with } \deg(r + sc) < \Phi\} = O(3^\Phi).$$

A further subtlety of our basic (i.e., without randomisation) addition algorithm is that there is not necessarily a unique τ-adic solution x to the equation $a + x = b$. For example, it is easy to check that $-1 + x = 1 + \tau$ has no solution in $\mathcal{T}_3$. Similarly, one can check that $-1 + x = -\tau + \tau^2$ has the two (necessarily equivalent) solutions $x = 1 - \tau + \tau^2$ and $x = -1$ when $\tau^2 - \tau + 2 = 0$. The extra randomisation in line 12 of Algorithm 1 is included precisely to 'smooth out' this issue. In particular, it greatly reduces the probability that an equation of the form $a + x = b$ cannot be solved in the important case when $\deg(a) < \deg(b)$.

4   GPS on Koblitz Curves with Fast Scalar Multiplication

The standard GPS scheme works for any abelian group for which the DLP is hard. In particular, one can use a Koblitz curve $E(\mathbb{F}_{2^m})$ and convert integers into Frobenius expansions to perform fast scalar multiplications. We briefly give the details: Fix $P \in E(\mathbb{F}_{2^m})$ of order r and generate the key pair $\{s, I = [-s]P\}$. We write $\tilde{x}$ for a τ-adic representation of an integer x. The prover picks a random integer r, converts it into τ-adic $\tilde{r}$, computes the commitment $X = [\tilde{r}]P$ and sends X to verifier. The verifier picks a random integer c and returns it to prover. The answer step is the same as the standard GPS (i.e., compute $y = r + sc \in \mathbb{Z}$) and the verification becomes checking that $X = [y]P + [c]I$ (which can be efficiently computed as $[\tilde{y}]P + [\tilde{c}]I$).

This method is very efficient, however there is the additional cost of converting an integer to a Frobenius expansion (plus the extra code footprint this requires). As noted by Solinas [19, 20], in any cryptographic protocol, instead of choosing a random integer and converting to a τ-adic expansion one can directly choose a τ-adic. This idea could be used for r in the standard GPS protocol, but one would still need to convert back to an integer for the computation $y = r + sc \in \mathbb{Z}$ in the online step. As mentioned earlier, the conversion algorithm requires modular arithmetic which is not otherwise needed as part of the GPS protocol. This results in additional overhead in running time and code on the device. We will see in the next section that these additional costs can be avoided if we use random τ-adics instead of random integers.

5 τ-GPS

When using τ-adic expansions instead of integers, we will get much faster computations for $I = [-s]P$, $X = [r]P$ and, in the verification step, for $X = [y]P + [c]I$. Figure 1 shows the τ-GPS scheme. We can repeat the protocol l times (though usually $l = 1$). We represent an element x picked at random from a set $\mathcal{X}$ by: $x \xleftarrow{r} \mathcal{X}$.

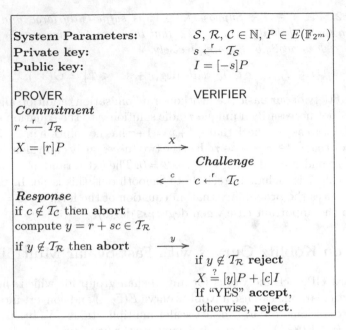

Fig. 1. τ-GPS

As with the original GPS protocol it is essential for the prover to perform a size check on the challenge c (otherwise, a dishonest verifier can send, for example, $c = \tau^{\mathcal{R}}$ and recover the prover's secret). The size check on y does not seem to be essential for security, but we include it to ease the security analysis. Heuristic 1 implies that the probability of aborting in the protocol is negligible.

Note that the computation of sc can be performed using the non-randomised version of Algorithm 1; the extra randomisation of the $\mathcal{K}'$ lowest order coefficients is only required when performing the addition with r. For further implementation details see Section 8.

6   Security Analysis

We closely follow [8] for our security analysis. In particular, we use the same security model as in [8]. We first consider attacks which recover the private key. Then, we analyse the sizes of the random τ-adic expansions such that the scheme is secure, i.e. the τ-GPS protocol really is a zero knowledge proof of the private key. Our analysis follows the approach of [7] and proves *completeness*, *zero knowledge* and *soundness*.

6.1   Discrete Logarithms

Given (P, I) it must be infeasible for an attacker to compute the private key s (similarly, given (P, X) to compute r). One could solve the DLP to get $\lambda \in \mathbb{N}$

such that $-I = [\lambda]P$ and then convert λ to a Frobenius expansion. Hence, we require the DLP in $\langle P \rangle$ to be hard. A standard choice is the Koblitz curve $E(\mathbb{F}_{2^{163}})$.

Alternatively, one could try to compute $s \in \mathcal{T}_S$ directly from (P, I). We call this the τ-DLP which we define below. This seems to be hard computational problem.

Definition 3. *The **Frobenius expansion discrete logarithm problem** (denoted τ-DLP) is: Given $P, Q \in E(\mathbb{F}_{2^m})$ and $\mathcal{S}$ to find $x \in \mathcal{T}_S$ (if it exists) such that $Q = [x]P$.*

For a full discussion of the τ-DLP, see [3]. Here, we merely remark that there is a variant of the Pollard kangaroo method which is expected to solve the τ-DLP in $O(2^{\frac{S}{2}+\epsilon})$ group operations, where the choice of ϵ is closely related to the probability of success in that method. As we have not verified in practice for large enough values of $\mathcal{S}$ yet, let us assume that, in the worst case, the method works well for $\epsilon = 0$. Hence we require $\mathcal{S} \geqslant 160$. As discussed in [3] there seems to be no loss of security by taking s to be in non-adjacent form, so we assume this in the below analysis.

6.2   Completeness

We want to show that GPS with τ-adic expansions is complete.

Theorem 1 (Completeness). *Suppose $\mathcal{R} \geqslant \Phi + \mathcal{K}$, for sufficiently large $\mathcal{K} \in \mathbb{N}$. Then, a prover who possess a valid key pair $(s, I = [-s]P)$ is accepted with overwhelming probability by a verifier.*

Proof. Clearly, the prover can compute $y = r + sc$. The verification step is:

$$[y]P + [c]I = [r + sc]P + [c][-s]P = [r]P + [sc]P + [-sc]P = [r]P = X,$$

which is successful. Finally, since $\mathcal{R} \geqslant \Phi + \mathcal{K}$, from Heuristic 1 we expect that $y \in \mathcal{T}_\mathcal{R}$ with overwhelming probability. $\qquad\qquad\square$

6.3   Size of r – The Zero Knowledge Proof

As with any public key identification or signature scheme it is important to show that runs of the protocol do not leak information about the private key. For the original GPS scheme it is proved in [8] that the protocol has statistical zero knowledge. Due to the strange properties of addition of Frobenius expansions using Algorithm 1 we seem to be unable to prove statistical zero knowledge and instead show computational zero knowledge with respect to a computational assumption.

First we remark that the τ-GPS protocol does leak information about s if one omits the extra randomisation step in the computation of $y = r + sc$. We give a brief sketch of the idea. Suppose a dishonest verifier inspects the constant coefficients of all polynomials, and always chooses c such that $c_0 = 1$. Then $y_0 = r_0 + s_0$ and if r_0 is uniformly distributed then the distribution of y_0 depends on the value of s_0. More precisely, if $s_0 = 1$ then the output distribution of y_0 in this case is 0 with probability $2/3$ and 1 with probability $1/3$; if $s_0 = 0$ then y_0 has uniform distribution in $\{-1, 0, 1\}$; if $s_0 = -1$ then $y_0 = 0$ with probability $2/3$ and -1 with probability $1/3$.

The above method can be extended to recover the first few coefficients of s. However, the randomisation of the first $\mathcal{K}'$ coefficients destroys the attack if $\mathcal{K}'$ is sufficiently large. Similarly, it seems hard to mount the attack on the $\mathcal{K}'$-th coefficient, since that is influenced by carry values propagating from addition of lower-degree terms.

We now state the computational assumption on which our security result relies.

Definition 4. *Let $C, S, \mathcal{K}$ and $\mathcal{K}'$ be parameters. Define $\Phi = S + C + 3$ and $\mathcal{R} = \max\{C + S + 3 + \mathcal{K}, m + \mathcal{K}'\}$. Let $s \in T_S$ and $I = [-s]P$. Let $\mathcal{M} = E(\mathbb{F}_{2^m}) \times T_C \times T_\mathcal{R}$. Let $f : E(\mathbb{F}_{2^m}) \to T_C$ be a function. Define the distribution on $\mathcal{M}$*

$$\mathcal{M}_{s,f,1} = \{(\alpha = [r]P, \beta = f(\alpha), \gamma = r + s\beta) : r \in T_\mathcal{R}\}$$

where r is selected from $T_\mathcal{R}$ uniformly at random and where the computation $r + s\beta$ is performed using Algorithm 1 with randomisation of the first $\mathcal{K}'$ coefficients. Define the distribution on $\mathcal{M}$

$$\mathcal{M}_{I,f,2} = \{(\alpha, \beta, \gamma) : \gamma \in T_{\Phi,\mathcal{R}}, \alpha = [\gamma]P + [\beta]I, \beta = f(\alpha)\}$$

where γ is selected uniformly at random.

We stress that these distributions are not the same. For example, in $\mathcal{M}_{I,f,2}$ there can be points $\alpha \in E(\mathbb{F}_{2^m})$ which are not of the form $[r]P$ for some $r \in T_\mathcal{R}$ (though they will typically be of the form $[r']P$ for some $r' \in T_{\mathcal{R}'}$ where $\mathcal{R}' - \mathcal{R}$ is small). More importantly, the distributions on γ are not the same in both cases: in the latter case γ is uniform in $T_{\Phi,\mathcal{R}}$ whereas due to the properties of addition using Algorithm 1 it is not clear whether the distribution of γ in the former case is close to uniform.

We now make a computational assumption regarding these distributions.

Assumption 1. *Let $\mathcal{A}$ be an algorithm running in polynomial time which is given I (but not s) and which samples elements from the distributions $\mathcal{M}_{s,f,1}$ and $\mathcal{M}_{I,f,2}$. Then we claim that $\mathcal{A}$ cannot distinguish the two distributions, namely that it does not have non-negligible advantage in being able to identify whether a given triple (α, β, γ) was drawn from $\mathcal{M}_{s,f,1}$ or $\mathcal{M}_{I,f,2}$.*

Theorem 2 (Zero Knowledge). *Let $C, S, \mathcal{K}$ and $\mathcal{K}'$ be security parameters and define $\Phi = C + S + 3$ and $\mathcal{R} = \max\{\Phi + \mathcal{K}, m + \mathcal{K}'\}$. Assume Heuristics 1, 2 and Assumption 1 above. Then the τ-GPS protocol has computational zero-knowledge if l and 3^C are polynomial and if $\mathcal{K}$ and $\mathcal{K}'$ are sufficiently large.*

Proof. We follow the proof of Theorem 2 of [8], but we cannot use exactly the same proof due to the strange properties of addition of Frobenius expansions using Algorithm 1. Let $\mathcal{A}_1$ be a dishonest verifier, who instead of picking challenges at random, chooses them based on previous iterations of the protocol, in order to try to obtain some knowledge about the private key. Following [8] we denote by $c(X, hist, \omega_\mathcal{A})$ the challenge chosen, depending on the commitment X, the history $hist$ of the protocol so far, and the random tape $\omega_\mathcal{A}$.

We now define an algorithm which simulates a round, using a random tape $\omega_\mathcal{A}$, without any knowledge of the secret s. The protocol is said to be computational zero knowledge if the (computationally bounded) adversary $\mathcal{A}$ cannot distinguish between the simulation and runs of the real protocol.

Step 1. Use $\omega_\mathcal{A}$ to choose random values $\bar{c} \in \mathcal{T}_\mathcal{C}$ and $\bar{y} \in \mathcal{T}_{\Phi,\mathcal{R}}$.
Step 2. Compute $\bar{X} = [\bar{y}]P + [\bar{c}]I$
Step 3. If $c(\bar{X}, hist, \omega_\mathcal{A}) \neq \bar{c}$ then return to Step 1, else return $(\bar{X}, \bar{c}, \bar{y})$.

We must show that, for any fixed random tape $\omega_\mathcal{A}$, it is computationally infeasible to distinguish the simulation from genuine runs of the protocol. This is exactly the computational assumption stated above, where $f(X)$ is the function $c(X, hist, \omega_\mathcal{A})$.

In other words, if the adversary $\mathcal{A}$ can distinguish between the simulation and the real protocol then it immediately solves the computational assumption. This completes the proof.   □

In practice we conjecture that $\mathcal{K} = 136$ and $\mathcal{K}' = 51$ are sufficient to obtain 80 bits of security. The motivation for this choice is as follows. First, $3^{\mathcal{K}'} \approx 2^{80}$, so the probability of guessing the randomness used in line 12 of Algorithm 1 is negligible. Second, $3^{\Phi+0.63\mathcal{K}-0.63}/3^{\Phi+\mathcal{K}} = 3^{0.37\mathcal{K}+0.63} \approx 2^{80}$, where $3^{\Phi+0.63\mathcal{K}-0.63}$ comes from Heuristic 1 and $3^{\Phi+\mathcal{K}} = 3^\mathcal{R}$ is the total number of possibilities for r in $\mathcal{T}_\mathcal{R}$, so the probability of learning anything about sc from $r + sc$ seems to be negligible.

6.4   Size of the Challenge c

The last task is to prove soundness of the scheme. We want to show that if the prover does not know the private key s, she can only be accepted by a verifier with very small probability. To do this we will show that if an adversary is accepted by a verifier with non-small probability, then she can recover the private key.

Theorem 3 (Soundness). *An adversary who does not know the private key is only accepted by a verifier with very small probability.*

Proof. Let us suppose that an adversary $\mathcal{A}$ is accepted with probability $\varepsilon > \frac{1}{3^c}$. Using the standard rewinding argument presented in Appendix B of [8] (inspired by [17]) we conclude that such an adversary $\mathcal{A}$ will be able to answer to two different challenges, c_1 and c_2 with y_1 and y_2, such that $[y_1]P + [c_1]I = X = [y_2]P + [c_2]I$.

So, we have $[(y_1 - y_2)]P = [(c_2 - c_1)]I$ and hence we can solve the DLP of I to the base P. Note that as we are working with elliptic curves the curve order is known, so there is no difficulty with solving this equation.   □

Therefore, we need $\mathcal{C}$ to be large enough such that the probability of cheating be very small. In practice, more than 2^{35} possibilities is enough to prevent an online attacker from guessing the right value for c. Since $\#\mathcal{T}_\mathcal{C} = 3^\mathcal{C}$ we need $\mathcal{C} \geqslant 23$.

7 Suggested Parameters

We have seen that one can take $m = 163$, $\mathcal{C} = 23$, $\mathcal{S} = 160$, $\mathcal{K} = 136$ and $\mathcal{K}' = 51$. This gives $\mathcal{R} = \max\{\Phi + \mathcal{K}, m + \mathcal{K}'\} = \max\{322, 214\} = 322$ which we summarise in Table 2. The first number in each column is the value of the parameter and the second number (between parenthesis) is the number of bits needed to represent an element of the set (using $\frac{5n}{3}$ bits to represent a τ-adic of length n and $n + 1$ bits if it can be chosen in non-adjacent form). We remark that, in principle, it should be possible to find better binary encodings for τ-adic expansions which require fewer than $\frac{5n}{3}$ bits.

Table 2. Numerical Example of GPS scheme with τ-adic expansions

$\mathcal{S}$	$\mathcal{C}$	$\mathcal{R}$
160 (161)	23 (38)	322 (537)

To speed up the computation of $s \times c$ we can use the Girault-Lefranc trick (see [10] for details). In other words, we insist that challenges c have at least $\mathcal{S} - 1$ zero coefficients between each pair of non-zero coefficients. It follows that computing $s \times c$ does not require Algorithm 1. This transforms the online step $y = r + sc$ into a single addition using Algorithm 1 with randomisation of the first $\mathcal{K}'$ coefficients.

In order to define $\mathcal{C}$ in this case we need to know how many τ-adic expansions with degree less than $\mathcal{C}$, hamming weight h and at least $\mathcal{S} - 1$ zero coefficients between each pair of non-zero coefficients exist. The proofs are elementary and ommitted due to lack of space.

Theorem 4. *The number of τ-adic expansions of degree less than $\mathcal{C}$ and with at least $\mathcal{S} - 1$ zero coefficients between each pair of non zero coefficients is*

$$Z_{\mathcal{C},\mathcal{S}} = \sum_{h=1}^{\lfloor \frac{\mathcal{C}+\mathcal{S}-1}{\mathcal{S}} \rfloor} 2^h \left[\binom{\mathcal{C} - h(\mathcal{S}-1)}{h} + \sum_{i=1}^{\mathcal{S}-1} \binom{\mathcal{C} - i - (h-1)(\mathcal{S}-1)}{h-1} \right]. \quad (2)$$

It follows that if $\mathcal{S} = 160$ then we need $\mathcal{C} = 797$ to get $Z_{\mathcal{C},\mathcal{S}} > 2^{35}$. This results in $\mathcal{R} = 1096$. Note that one can transmit/store c using much fewer than 729 bits. However, r and y are not sparse, so would need around 1825 bits to represent.

Table 3. Parameters for Girault-Lefranc variant τ−GPS

$\mathcal{S}$	$\mathcal{C}$	$\mathcal{R}$
160 (161)	797 (798)	1096 (1827)

8 Performance Analysis

We give more detail about how to efficiently compute each of the three steps of the GPS protocol.

Step 1: We are supposed to choose a random element $r \in \mathcal{T}_\mathcal{R}$ (where $\mathcal{R}$ may be 322 or 1096) and compute $X = [r]P$. Note that $\mathcal{R}$ is larger than m. In practice it would be more efficient to reduce r modulo $\tau^m - 1$ before performing the computation of X (this may lead to coefficients outside the set $\{-1, 0, 1\}$).

A more efficient alternative is as follows. Choose a random $r' \in \mathcal{T}_m$ and compute $X = [r']P$. It is only necessary to store r', requiring $\approx 5m/3$ bits. Indeed, one could choose r' in non-adjacent form, in which case at most $m + 1$ bits are required.

When the long value r is needed in the online step we can simply choose the coefficients r_j randomly subject to the constraint that for $0 \leqslant i < m$ we have

$$\sum_{j=0}^{\lfloor (n-i)/m \rfloor} r_{jm+i} = r'_i$$

which can be done efficiently. Note that this variant is not covered by our security analysis since not all possible values of $r \in \mathcal{T}_\mathcal{R}$ can arise. A security analysis of this case is a topic for future research.

Step 2: The computation of $r + sc$ is done using repeated calls to Algorithm 1. One could also use the Karatsuba idea to speed-up the polynomial multiplication. For more details of improvements see [10].

Step 3: To compute $[y]P + [c]I$ one can reduce y (and c if necessary) modulo $\tau^m - 1$. One can also use the standard multiexponentiation and precomputation techniques to speed this up (see [1]).

We now briefly compare τ-GPS with the elliptic curve GPS variant of Section 4. The standard GPS has lower bandwidth than τ-GPS and the online computations are simpler and we expect them to be faster for standard GPS. The offline computations are quite efficient in both cases (especially when compared with GPS over an RSA modulus). A further advantage of standard GPS is that the security assurance is statistical zero knowledge rather than computational zero knowledge. The advantage of τ-GPS is that it does not require conversion between integers and Frobenius expansions.

We now list some extensions and avenues for future research. See [2] for further details.

- As usual, one can obtain a signature scheme from the τ-GPS scheme using the Fiat-Shamir heuristic (see [8]). This gives rise to an online/offline signature scheme which is typically more efficient than the schemes of Shamir and Tauman [18].
- One can generalise to elliptic curves over larger fields $\mathbb{F}_p$, in which case the coefficient set is not $\{-1, 0, 1\}$ but $\{-(p-1)/2, \ldots, -1, 0, 1, \ldots, (p-1)/2\}$.
- In the case of characteristic greater than two one could use Edwards elliptic curves [4].
- One can use an intermediate between Girault-Lefranc and standard GPS for computing $s \times c$ efficiently for smaller $\mathcal{C}$.
- Finding more efficient bit representations for τ-adic expansions (possibly by working with elliptic curves over fields larger than $\mathbb{F}_2$) is a topic for future research.

9   Conclusion

We presented the GPS identification scheme using τ-adic expansions. This speeds up the running time of the offline steps compared with standard GPS. The paper shows that cryptographic protocols can be made to operate with Frobenius expansions instead of integers, and this idea may have wider applications.

We now give our recommendations on efficient identification protocols for constrained devices. First, since the resources on the device are limited we recommend using Koblitz elliptic curves (rather than RSA moduli as proposed in [9, 10]). If there are no constraints on the offline computation time then we suggest using the standard GPS protocol or Girault-Lefranc. If the offline computation time is also limited then it is natural to improve the performance of the offline steps using Frobenius expansions. The precise choice of protocol then depends on the application:

- If bandwidth is the most precious resource then we recommend Schnorr signatures.
- If computation time of the online stage is the most precious resource then we recommend the Girault-Lefranc method [10].
- If both bandwidth and computation time are precious then we recommend the standard GPS scheme [8].
- If code area and/or power consumption are the most precious resources then we recommend τ-GPS.

References

1. Avanzi, R., Cohen, H., Doche, C., Frey, G., Lange, T., Nguyen, K., Vercauteren, F.: Handbook of elliptic and hyperelliptic curve cryptography. Discrete Mathematics and its Applications. Chapman & Hall/CRC, Boca Raton (2006)
2. Benits, W.: Applications of Frobenius expansions in elliptic curve cryptography, PhD thesis in preparation

3. Benits, W., Galbraith, S.: The Frobenius expansion DLP, preprint
4. Bernstein, D.J., Lange, T.: Faster addition and doubling on elliptic curves. In: Kurosawa, K. (ed.) ASIACRYPT 2007. LNCS, vol. 4833. pp. 29–50. Springer, Heidelberg (2007)
5. Bosma, W., Cannon, J., Playoust, C.: The MAGMA algebra system I: the user language. Journal of Symbolic Computation 24, 235–265 (1997)
6. Ebeid, N., Hasan, M.A.: On τ-adic representations of integers. Designs, Codes and Cryptography 45(3), 271–296 (2007)
7. Feige, U., Fiat, A., Shamir, A.: Zero-knowledge proofs of identity. Journal of Cryptology 1(2), 77–94 (1988)
8. Girault, M., Poupard, G., Stern, J.: On the fly authentication and signature schemes based on groups of unknown order. J. Crypt. 19(4), 463–487 (2006)
9. Girault, M.: Self-certified public keys. In: Davies, D.W. (ed.) EUROCRYPT 1991. LNCS, vol. 547, pp. 490–497. Springer, Heidelberg (1991)
10. Girault, M., Lefranc, D.: Public key authentication with one (online) single addition. In: Joye, M., et al. (eds.) CHES 2004. LNCS, vol. 3156. pp. 413–427. Springer, Heidelberg (2004)
11. Koblitz, N.: CM-curves with good cryptographic properties. In: Feigenbaum, J. (ed.) CRYPTO 1991. LNCS, vol. 576. pp. 279–287. Springer, Heidelberg (1992)
12. Müller, V.: Fast multiplication on elliptic curves over small fields of characteristic two. Journal of Cryptology 11(4), 219–234 (1998)
13. Okamoto, T., Katsuno, H., Okamoto, E.: A fast signature scheme based on new on-line computation. In: Boyd, C., Mao, W. (eds.) Information Security. LNCS, vol. 2581. pp. 111–121. Springer, Heidelberg (2003)
14. Poupard, G., Stern, J.: Security analysis of a practical "on the fly" authentication and signature generation. In: Nyberg, K. (ed.) EUROCRYPT 1998. LNCS, vol. 1403. pp. 422–436. Springer, Heidelberg (1998)
15. Rivest, R.L., Cormen, T.H., Leiserson, C.E., Stein, C.: Introduction to algorithms, 2nd edn. MIT Press and McGraw-Hill (2001)
16. Schnorr, C.P.: Efficient identification and signatures for smart cards. In: Brassard, G. (ed.) CRYPTO 1989. LNCS, vol. 435. pp. 239–252. Springer, Heidelberg (1990)
17. Schnorr, C.P.: Efficient signature generation by smart cards. Journal of Cryptology 4(3), 161–174 (1991)
18. Shamir, A., Tauman, Y.: Improved online/offline signature schemes. In: Kilian, J. (ed.) CRYPTO 2001. LNCS, vol. 2139. pp. 355–367. Springer, Heidelberg (2001)
19. Solinas, J.A.: An improved algorithm for arithmetic on a family of elliptic curves. In: Kaliski Jr., B.S. (ed.) CRYPTO 1997. LNCS, vol. 1294, pp. 357–371. Springer, Heidelberg (1997)
20. Solinas, J.A.: Efficient arithmetic on Koblitz curves. Des. Codes Cryptography 19(2-3), 195–249 (2000)

Searching for Messages Conforming to Arbitrary Sets of Conditions in SHA-256

Marko Hölbl[1], Christian Rechberger[2], and Tatjana Welzer[1]

[1] Faculty of Electrical Engineering and Computer Science, University of Maribor,
Maribor, Slovenia
marko.holbl@uni-mb.si
http://www.feri.uni-mb.si
[2] Institute of Applied Information Processing and Communications (IAIK), Graz
University of Technology, Graz, Austria
http://www.iaik.tugraz.at/research/krypto

Abstract. Recent progress in hash functions analysis has led to collisions on reduced versions of SHA-256. As in other hash functions, differential collision search methods of SHA-256 can be described by means of conditions on and between state and message bits. We describe a tool for efficient automatic searching of message pairs conforming to useful sets of conditions, *i. e.* stemming from (interleaved) local collisions. We not only considerably improve upon previous work [7], but also show the extendability of our approach to larger sets of conditions.

Furthermore, we present the performance results of an actual implementation and pose an open problem in this context.

Keywords: hash function, SHA-256, conditions, differential collision search.

1 Introduction

Owing to the recent cryptanalytic results on MD5 [11], SHA-1 [1,9,10] and similar hash functions, the influence of these attacks on members of the SHA-2 family (i.e. SHA-224, SHA-256, SHA-384 and SHA-512) [8] is an important issue.

Although SHA-256 is considered as the successor of SHA-1 [8] and hence a very important cryptanalytic target, it received comparatively little analysis in from the open cryptographic community [4,5,6,7,12].

Review of previous work. In [7] Mendel et al. presented message pairs for a collision of SHA-256 reduced to 18 steps and a complex collision characteristic covering 19 steps. Their attacks is an extension of the analysis by Chabaud and Joux [2], and Rijmen and Oswald [9]. The 18-step collision in [7] is an example of using a single local collision. The basic idea of local collisions is to cancel out differences as soon as possible after they are introduced. All collision search attacks on members of the SHA-2 family (SHA-224, SHA-256, SHA-384, SHA-512) as well as its predecessors (SHA, SHA-1, to some extent also MD4, MD5 and others) use local collisions as their basic building block. The 19-step characteristic is an example of how to interleave several local collision.

S. Lucks, A.-R. Sadeghi, and C. Wolf (Eds.): WEWoRC 2007, LNCS 4945, pp. 28–38, 2008.

Discussion of new results. The step transformations employed by all these hash functions prevent deterministic propagations of differences. However it is possible to explicitly write out the equations that need to be met such that differences propagate as expected. Subsequently these equations will be refereed to as sufficient conditions.

Using the set of sufficient conditions that describe local collisions in SHA-256, we develop a tool for automatic searching of such messages that conform to these conditions. In this paper we present details about the tool and give performance results of efficient searching of messages conforming to large sets of condition for SHA-256, i. e. several interleaved local collisions simultaneously. In fact the method is general enough to allow to find a pair of conforming messages for any consistent set of conditions.

Since many applications did already migrate from SHA-1 to SHA-256, it is vital to have information on its resistance against various cryptanalytic methods. The methods presented in this paper will serve as a building block to estimate the workload for collision attacks on SHA-256 (and all other members of the SHA-2 family for that matter) following newly devised differential characteristics.

Organization of this paper. We give a short review of SHA-256 in Section 2. Section 3 describes the algorithm in detail. Additionally, it presents details on the outline of the algorithm (Section 3.3), the data structure used to store the set of sufficient conditions (section 3.1), 'intelligent' sorting of sufficient conditions (Section 3.2), details on parts of the algorithm covering direct message modification (Section 3.3), modification over the Σ function (Section 3.3) and the correction approach in case of 'special' sufficient conditions (Section 3.3). During our development, we came across several problems part of which we resolved. The details are described in Section 3.4. Furthermore, performance measurements and comparison with previous results (in [7]) regarding message searching is given in Section 4. We conclude the paper in Section 5.

2   Review of SHA-256

The following notation will be used through the whole paper (Table 1). We only give a brief description of SHA-256, to the extent needed for understanding the paper. A complete description of SHA-256 can be found in [8]. SHA-256 is an iterated cryptographic hash function based on a compression function that updates the eight 32-bit state variables $A, \ldots, H$ according to the values of 16 32-bit words $M_0, \ldots, M_{15}$ of the message. The compression function consists of 64 identical steps as presented in Fig. 1. The step transformation employs the bitwise Boolean functions f_{MAJ} and f_{IF}, and two GF(2)-linear functions

$$\Sigma_0(x) = ROTR^2(x) \oplus ROTR^{13}(x) \oplus ROTR^{22}(x),$$
$$\Sigma_1(x) = ROTR^6(x) \oplus ROTR^{11}(x) \oplus ROTR^{25}(x).$$

The i-th step uses a fixed constant K_i and the i-th word W_i of the expanded message. The message expansion works as follows. An input message is padded

Table 1. Notation

notation	description
$A_i \ldots H_i$	state variables at step i of the compression function
$A \oplus B$	bit-wise XOR of state variable A and B
$A + B$	addition of state variable A and B modulo 2^{32}
$ROTR^n(A)$	bit-rotation of A by n positions to the right
$SHR^n(A)$	bit-shift of A by n positions to the right
N	number of steps of the compression function

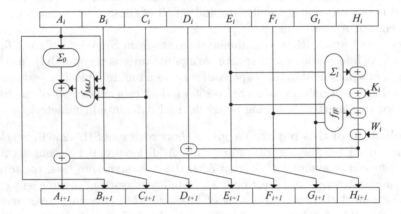

Fig. 1. One step of the state update transformation of SHA-256

and split into 512-bit message blocks. Let ME denote the message expansion function. ME takes as input a vector M with 16 coordinates and outputs a vector W with N coordinates. The coordinates W_i of the expanded vector are generated from the initial message M according to the following formula:

$$W_i = \begin{cases} M_i & \text{for } 0 \leq i < 16 \\ \sigma_1(W_{i-2}) + W_{i-7} + \sigma_0(W_{i-15}) + W_{i-16} & \text{for } 16 \leq i < N \end{cases}$$

Taking a value for N different to 64 results in a step-reduced (or extended) variant of the hash function. The functions $\sigma_0(x)$ and $\sigma_1(x)$ are defined as follows: $\sigma_0(x) = ROTR^7(x) \oplus ROTR^{18}(x) \oplus SHR^3(x)$ and $\sigma_1(x) = ROTR^{17}(x) \oplus ROTR^{19}(x) \oplus SHR^{10}(x)$.

3   The Message Search

According to Mendel et al. a 9-step local collision for SHA-256 is possible [7]. Furthermore, they give a characteristic (in Table 2) from which a set of conditions can be derived (Table 3 from [3]). Employing these conditions we develop a tool for automatic searching of messages conforming the set of sufficient condition.

Table 2. Example of a local collision for SHA-256 [7]

Step	W'	A'	B'	C'	D'	E'	F'	G'	H'	HW
01	80000000	80000000	0	0	0	80000000	0	0	0	2
02	22140240	0	80000000	0	0	20040200	80000000	0	0	5
03	42851098	0	0	80000000	0	0	20040200	80000000	0	5
04	0	0	0	0	80000000	0	0	20040200	80000000	5
05	80000000	0	0	0	0	80000000	0	0	20040200	4
06	22140240	0	0	0	0	0	80000000	0	0	1
07	0	0	0	0	0	0	0	80000000	0	1
08	0	0	0	0	0	0	0	0	80000000	1
09	80000000	0	0	0	0	0	0	0	0	0

Table 3. Sufficient conditions for the SHA-256 local collision given in Table 2 [3]

no	type	condition
01	hard	$E_{04,09} = 0$
02	hard	$E_{04,18} = 0$
03	hard	$E_{04,29} = 0$
04	hard	$E_{04,31} = 1$
05	hard	$E_{05,09} = 1$
06	hard	$E_{05,18} = 1$
07	hard	$E_{05,29} = 1$
08	hard	$E_{07,31} = 0$
09	hard	$E_{08,31} = 1$
10	hard	$Z_{02,06} \oplus W_{02,06} = 1$
11	hard	$E_{03,09} \oplus W_{02,09} = 0$
12	hard	$Y_{02,09} \oplus W_{02,09} = 1$
13	hard	$E_{03,18} \oplus W_{02,18} = 0$
14	hard	$Y_{02,18} \oplus W_{02,18} = 1$
15	hard	$Z_{02,20} \oplus W_{02,20} = 1$
16	hard	$Z_{02,25} \oplus W_{02,25} = 1$
17	hard	$E_{03,29} \oplus W_{02,29} = 0$
18	hard	$Y_{02,29} \oplus W_{02,29} = 1$
19	hard	$A_{01,31} \oplus A_{00,31} = 0$
20	hard	$E_{01,31} \oplus E_{00,31} = 0$
21	hard	$Z_{03,03} \oplus W_{03,03} = 1$
22	hard	$Z_{03,04} \oplus W_{03,04} = 1$
23	hard	$Z_{03,07} \oplus W_{03,07} = 1$
24	hard	$E_{02,09} \oplus E_{01,09} = 0$
25	hard	$Z_{03,12} \oplus W_{03,12} = 1$
26	hard	$Z_{03,16} \oplus W_{03,16} = 1$
27	hard	$E_{02,18} \oplus E_{01,18} = 0$
28	hard	$Z_{03,18} \oplus W_{03,18} = 1$
29	hard	$Z_{03,23} \oplus W_{03,23} = 1$
30	hard	$Z_{03,25} \oplus W_{03,25} = 1$
31	hard	$E_{02,29} \oplus E_{01,29} = 0$
32	hard	$A_{03,31} \oplus A_{01,31} = 0$
33	hard	$A_{04,31} \oplus A_{03,31} = 0$
34	hard	$Z_{06,06} \oplus W_{06,06} = 1$
35	hard	$Z_{06,20} \oplus W_{06,20} = 1$
36	hard	$Z_{06,25} \oplus W_{06,25} = 1$
37	hard	$E_{05,31} \oplus E_{04,31} = 0$
38	hard	$E_{03,31} \oplus Z_{03,30} \oplus W_{03,30} = 1$
39	easy	$W_{02,09} \oplus W_{06,09} = 1$
40	easy	$W_{02,18} \oplus W_{06,18} = 1$
41	easy	$W_{02,29} \oplus W_{06,29} = 1$

An important new aspect of our work is that we employ a particular *order* in which conditions are considered. In this section the data structure for sorting the sufficient conditions, the prerequisites for the tool (sorting of sufficient condition) and details about the search algorithm are presented.

Table 4. The matrix data structure for the sufficient conditions

index	saved value
1	sufficient condition 1 - message word
2	sufficient condition 1 - step
3	sufficient condition 1 - bit
4	sufficient condition 2 - message word
5	sufficient condition 2 - step
6	sufficient condition 2 - bit
7	sufficient condition 3 - message word
8	sufficient condition 3 - step
9	sufficient condition 3 - bit
10	sufficient condition 4 - message word
11	sufficient condition 4 - step
12	sufficient condition 4 - bit
13	sufficient condition 5 - message word
14	sufficient condition 5 - step
15	sufficient condition 5 - bit
15	right side value

3.1   The Data Structure for Storing Sufficient Conditions

In Table 3 Z and Y denote the outputs of $\Sigma_0(x)$ and $\Sigma_1(x)$. Z and Y can be written in expanded form as follows:

$$Z_{i,j} = A_{i,j+2} \oplus A_{i,j+13} \oplus A_{i,j+22},$$
$$Y_{i,j} = E_{i,j+6} \oplus E_{i,j+11} \oplus E_{i,j+25}.$$

The expanded form of sufficient conditions is used by the search tool which makes the search easier.

After the sufficient conditions have been expanded, we defined a matrix data structure for storing the sufficient conditions. The matrix is of size $16 \times nrSC$, where $nrSC$ is the number of sufficient conditions. The line of a matrix is of length 16, because of potential maximum of 5 parts of a sufficient condition:

e.g. $E_{03,31} \oplus Z_{03,30} \oplus W_{03,30} = 1$ is expanded to
$E_{03,31} \oplus E_{03,05} \oplus E_{03,10} \oplus E_{03,24} \oplus W_{03,30} = 1$.

Each part of the sufficient condition (which consist of message word, step and bit) is stored in one array element. The total of 16 is calculated due the the following formula (maximum number of sufficient conditions is 5): 5 sufficient conditions × 3 array elements for storing one sufficient condition + right side value = 16. For better readability the matrix structure is depicted in Table 4.

3.2   Sorting of Conditions

The order with which conditions are dealt with has an impact on the runtime of the algorithm we are about to propose. Here we describe the criteria that are useful to determine a good order for the conditions. The priorities of the rules bellow are defined by their sequence numbers (1 is applied first, 2 second, etc.).

1. Sufficient conditions are sorted in such a way, that parts with larger step numbers are sorted before parts with lower ones: $A_{i,j} \oplus A_{i+1,j} = v$ is transformed to $A_{i+1,j} \oplus A_{i,j} = v$ due to the fact that $A_{i+1,j}$ includes step $i + 1$ and $A_{i,j}$ step i. This must be done for the whole set of sufficient conditions.
2. Sufficient conditions are sorted together as to the step of their first part: $E_{i+1,j} \oplus E_{i,j} = v$ and $E_{i+1,j} \oplus E_{i,j+1} = v$ are sorted together regarding they both apply for step $i+1$ (Nevertheless that they also include a part applying for the step i).
3. Shorter sufficient conditions are sorted before longer ones. Namely, shorter sufficient condition include less degrees a freedom regarding the modification possibilities: $E_{i,j} \oplus W_{i,j} = v$ is sorted before $E_{i,j} \oplus Z_{i,j} \oplus W_{i,j} = v$.
4. Sufficient conditions consisting of parts including state update variables E are sorted before those including A. *Example:* $E_{i,j} \oplus W_{i,j} = v$ is sorted before $A_{i,j} \oplus A_{i,y} = v$.

3.3   The Search Algorithm

After the sufficient conditions have been transformed into the form described in 3.1 and sorted according to the rules in section 3.2, we execute the search algorithm. It includes several steps which are summarized below and further explained into detail in the sequel sub-sections.

check current sufficient condition SC_k
if (current SC_k not fulfilled)
  if (the current part of SC_k includes $W_{i,j}$)
    correct appropriate bit in $W_{i,j}$
  if (current part of SC_k includes $A_{i,j}$ or $E_{i,j}$)
    correct appropriate bit in $W_{i-1,j}$
  if (direct correction is not possible (modification of $W_{i-1,j}$ resp $W_{i,j}$)
    if (current part of SC_k includes $A_{i,j}$)
      correct over Σ_0 by correcting W_{i-2} (2 steps back)
      or
      correct the appropriate register of the IV
    if (current part of SC_k includes $E_{i,j}$)
      correct over Σ_1 by correcting W_{i-2} (2 steps back)
      or
      correct the appropriate register of the IV
  if ('special' sufficient condition)
    correct W_{i-5} by going 5 steps back
    or
    correct the appropriate register of the IV
  else
    generate a new random message and start from the beginning

As can be seen from above, we firstly try to correct messages directly in order to fulfill a specific sufficient condition. In case this is not possible, we try indirect corrections over Σ function. Nevertheless, we still correct message word W, but we apply corrections to different bits and have to go two steps back (W_{i-2}). When we are dealing with so-called 'special' sufficient conditions, we have an additional possibility to change W 5 steps back (i.e. W_{i-5}). If none of the above corrections can be applied, we have to generate a new random messages and start the algorithm from the beginning.

Direct corrections. Direct correction of messages is applied to every part of a particular sufficient conditions SC_k. In case of sufficient condition $E_{i,j} \oplus E_{i,j+1} \oplus E_{i,j+2} \oplus E_{i,j+3} \oplus W_{i,j+4} = v$, the algorithm tries to correct $W_{i-1,j}$, $W_{i-1,j+1}$, $W_{i-1,j+2}$, $W_{i-1,j+3}$ or $W_{i-1,j+4}$. When flipping the appropriate bit, it checks if the the flip influences any previous conditions $SC_1 \ldots SC_{i-1}$. If so, the flip is undone. This is only true if we are correcting A or E parts of sufficient conditions, because we have to go one step back. When dealing with part of sufficient conditions which include message word W_i (e.g. $W_{i,j+4}$), the check is only done in cases where $i < step$, because we are flipping bits of messages word applying to previous steps. In case where $i = step$ checking is redundant, because W_i only influence successive steps.

```
if (part of SC_k is of type W_{i,j})
  if (i = step)
    correct W_{i,j} without checking
  if (i < step)
    if (bit flip in W_{i,j} influences SC_1 ... SC_{i-1})
      undo bit flip
    else
      correct W_{i,j}
if (part of SC_k is of type A_{i,j} or E_{i,j})
  if(bit flip in W_{i-1,j} influences SC_1 ... SC_i)
    undo bit flip
  else
    correct W_{i-1,j}
```

Corrections over SIGMA. In case where no direct correction is possible, corrections over Σ functions are employed. If we wanted to correct $E_{i,j}$ over Σ_1, the following messages word are potentially corrected: $W_{i-2,j+6}$, $W_{i-2,j+11}$ or $W_{i-2,j+25}$. As the bit flip is applied, the tool checks if the bit flip influences any previous conditions for steps $1 \ldots s$. If so, the flip is undone. In case of A parts, the corrections is done similarly, except for the following change: messages word that corrected are $W_{i-2,j+2}$, $W_{i-2,j+13}$ or $W_{i-2,j+22}$ (because function Σ_0 is employed).

```
if (current SC_k part = A part)
  if (bit flip in W_{i-2,j+2}, W_{i-2,j+13} or W_{i-2,j+22} influence SC_1 ... SC_i)
    do not correct
```

else
  correct either $W_{i-2,j+2}$, $W_{i-2,j+13}$ or $W_{i-2,j+22}$
if(current SC_k part $= E$ part)
 if (bit flip in $W_{i-2,j+6}$, $W_{i-2,j+11}$ or $W_{i-2,j+25}$ influence $SC_1 \ldots SC_i$)
 do not correct
 else
  correct either $W_{i-2,j+6}$, $W_{i-2,j+11}$ or $W_{i-2,j+25}$

Corrections of 'special' sufficient conditions. When dealing with 'special' sufficient conditions, we apply an additional correcting approach besides the once mentioned earlier. In case of an E part sufficient condition, corrections are done 5 steps back. When this is not possible, the corresponding register in the initialization vector (IV) is corrected - i.e. with regard to the current step we correct either register A, B, C or D. Registers of the IV are corrected as follows:

– register A in the IV is corrected when $step = 4$,
– register B in the IV is corrected when $step = 3$,
– register C in the IV is corrected when $step = 2$,
– register D in the IV is corrected when $step = 1$,

When the step $i \geq 5$, corrections on $W_{i-5,j}$ are employed. As in the previous cases the effect of bit flip on conditions for steps $1 \ldots s - 4$ must be checked. In case of an A part 'special' sufficient condition, we flip bits in $W_{i-1,j}$. Additionally, influences on conditions for steps $1 \ldots s$ have to be checked.

if (A part)
 if (bit flip in $W_{i-1,j}$ influence $SC_1 \ldots SC_i$)
 do not flip bit
 else
  correct $W_{i-1,j}$
if (E part)
 if ($1 \leq step \leq 4$)
 correct D, C, B or A in the IV
 if ($step \geq 5$)
  if (bit flip in $W_{i-1,j}$ has influence on conditions for steps $1 \ldots s - 4$)
  do not flip bit
  else
   correct $W_{i-5,j}$

3.4 Problems Encountered during Development of the Tool

Firstly, SHA-256 employs a complex step update operation (Fig. 1). It simultaneously updates two state variables, thus makes message search difficult. Hence, we have to consider that message word W_i influences both state variables A_i and E_i.

Secondly, particular sufficient conditions interfere with each other. A specific bit of a state variable is included in multiple sufficient conditions in the same step.

An example two such conditions are sufficient conditions $A_{01,31} \oplus A_{00,31} = 0$ and $E_{01,31} \oplus E_{00,31} = 0$. This limits the capabilities for bit changing through message word modification. We tried to solve this problem by checking the influence of bit flips on previous sufficient conditions which were already fulfilled. An exception are bit overlaps of message words W_i and state variables A_i resp. E_i. In this case, the specific bits are changed, because W_i has influence on the successive step.

Thirdly, some sufficient conditions include not only state variable conditions but also message word conditions which influence sufficient conditions in successive steps. An example is sufficient condition $Z_{02,06} \oplus W_{02,06} = 1$. When making message modification of W_i in successive steps we also have to consider these. In case we change the message word W_i (step i), we would have to check the sufficient conditions in the previous step (namely $i - 1$). Thus, this limits the message modification possibilities.

Fourthly, there are 'special' types of sufficient conditions, e.g. ($E_{03,09} \oplus W_{02,09} = 0$). This type of conditions include a state variable from the current (e.g. A_i resp. E_i) step and a message word $W_i - 1$ from the previous step. In this case it is impossible to modify message word from step $i - 1$ in order to flip a bit in step i. In such cases we try to correct over Σ functions and by going 5 steps back (as described in section 3.3.

In some special cases bit carries play an important role. Nevertheless bit i is changed in W_i, which should influence the same bit in the state variable of step $i + 1$, the flip also influences other bits. We solve such problems by checking influences of bit flips on sufficient conditions and undo them if necessary.

4   Performance

The tool was tested on 4 sets of conditions corresponding to $1 \ldots 4$ local collisions. It should be noticed, that so-called easy condition (conditions including only message words W and no state update variables A or E) were not counted in the performance results as they are easy to fulfill on beforehand. Therefore the actual number of sufficient conditions ranges from $38 \ldots 92$. We derived these sets from the original set of 41 by Mendel et al. ([7]). The performance measurements are summarized in table 5.

For smaller sets, the time taken to find a message is low (5 ms). As the number of sufficient conditions gets larger the time complexity raises very fast. For the largest set of conditions, the time complexity is approximately 2, 5 minutes. Similarly, the number of randomly generated messages increases from 7 to 183.323 and the number of SHA-256 step operations needed to find a message from 2.299 for 40 sufficient conditions to 1.956.071 for 100. The number of randomly generated messages refers to the number of messages which had to be generated if we start from the beginning (if not other correction was possible). For further details the reader should refer to Section 3.3.

In comparison with previous results by Mendel et al. [7], we observe a large improvement of searching speed. Furthermore our results are first results on multiple

Table 5. Performance measurements

	our results				results of [7]
number of sufficient conditions	38	56	74	92	38
number of randomly generated messages	7	200	2.211	183.323	/
number of SHA-256 step operations	2.229	60.782	922.780	1.956.071	/
time taken [s]	$0,005$	$0,151$	$1,988$	$154,266$	$0,5$

sets of conditions for SHA-256 and open up possibilities of searching messages conforming to even larger sets of conditions.

5   Conclusion and Outlook

We presented the problems of finding message pairs conforming to simple characteristics for SHA-256 and their respective conditions. SHA-256 is the only hash function within the MD4 family of hash functions in which this problem seems to be hard even in the first steps. As a main reason, we identified cyclical dependencies between condition due to the fact that two chaining variables are updated in every step instead of only one in all previous designs.

We significantly improve upon previous work [7] by 1) tracking of bit flips backwards through building blocks like Σ and 2) intelligent sorting of conditions. The results still indicate an exponential growth in runtime(albeit with a now much lower constant than before). It is an open question if practical methods with non-exponential runtime can be found.

Acknowledgements. We would like to thanks the anonymous reviewers for their helpful remarks and questions.

References

1. Biham, E., Chen, R., Joux, A., Carribault, P., Lemuet, C., Jalby, W.: Collisions of SHA-0 and Reduced SHA-1. In: Cramer, R. (ed.) EUROCRYPT 2005. LNCS, vol. 3494, pp. 36–57. Springer, Heidelberg (2005)
2. Chabaud, F., Joux, A.: Differential Collisions in SHA-0. In: Krawczyk, H. (ed.) CRYPTO 1998. LNCS, vol. 1462, pp. 56–71. Springer, Heidelberg (1998)
3. De Cannière, C., Mendel, F., Pramstaller,N., Rechberger, C., Rijmen, V.: SHA Evaluation Report for CRYPTREC, January 21 (2006)
4. Gilbert, H., Handschuh, H.: Security analysis of SHA-256 and sisters. In: Matsui, M., Zuccherato, R. (eds.) SAC 2003. LNCS, vol. 3006. pp. 175–193. Springer, Heidelberg (2003)
5. Hawkes, P., Paddon, M., Rose, G.G.: On corrective patterns for the SHA-2 family. Cryptology ePrint Archive, Report /207 (August 2004), http://eprint.iacr.org/

6. Matusiewicz, K., Pieprzyk, J., Pramstaller, N., Rechberger, C., Rijmen, V.: Analysis of simplified variants of SHA-256. In: Proceedings of WEWoRC 2005, LNI P-74, pp. 123–134 (2005)
7. Mendel, F., Pramstaller, N., Rechberger, C., Rijmen, V.: Analysis of Step-Reduced SHA-256. In: Robshaw, M. (ed.) FSE 2006. LNCS, vol. 4047. pp. 126–143. Springer, Heidelberg (2006)
8. National Institute of Standards and Technology (NIST). FIPS-180-2: Secure Hash Standard (August 2002), http://www.itl.nist.gov/fipspubs/
9. Rijmen, V., Oswald, E.: Update on SHA-1. In: Menezes, A. (ed.) CT-RSA 2005. LNCS, vol. 3376. pp. 58–71. Springer, Heidelberg (2005)
10. Wang, X., Yin, Y.L., Yu, H.: Finding Collisions in the Full SHA-1. In: Shoup, V. (ed.) CRYPTO 2005. LNCS, vol. 3621. pp. 17–36. Springer, Heidelberg (2005)
11. Wang, X., Yu, H.: How to Break MD5 and Other Hash Functions. In: Cramer, R. (ed.) EUROCRYPT 2005. LNCS, vol. 3494. pp. 19–35. Springer, Heidelberg (2005)
12. Yoshida, H., Biryukov, A.: Analysis of a SHA-256 variant. In: Preneel, B., Tavares, S. (eds.) SAC 2005. LNCS, vol. 3897. pp. 245–260. Springer, Heidelberg (2006)

Efficient Hash Collision Search Strategies on Special-Purpose Hardware

Tim Güneysu, Christof Paar, and Sven Schäge

Horst Görtz Institute for IT Security, Ruhr University Bochum, Germany
{gueneysu, cpaar}@crypto.rub.de, sven.schaege@nds.rub.de

Abstract. Hash functions play an important role in various cryptographic applications. Modern cryptography relies on a few but supposedly well analyzed hash functions which are mostly members of the so-called MD4-family. This work shows whether it is possible to significantly speedup collision search for MD4-family hash functions using special-purpose hardware. A thorough analysis of the computational requirements for MD4-family hash functions and corresponding collision attacks reveals that a microprocessor based architecture is best suited for the implementation of collision search algorithms. Consequently, we designed and implemented a (concerning MD4-family hash-functions) general-purpose microprocessor with minimal area requirements and, based on this, a full collision search unit. Comparing the performance characteristics of both ASICs with standard PC processors and clusters, it turns out that our design, massively parallelized, is nearly four times more cost-efficient than parallelized standard PCs. Although with further optimizations this factor can certainly be improved, we believe that special-purpose hardware does not provide a too significant benefit for hash collision search algorithms with respect to modern off-the-shelf general-purpose processors.

Keywords: Hash functions, Special-purpose Hardware, Crypto Attacks.

1 Introduction

Many basic and complex cryptographic applications make extensive use of cryptographic hash functions. They offer valuable security properties and computational efficiency. In combination, these features are particularly interesting for accelerating asymmetric cryptographic protocols. Usually, the security of a cryptographic protocol is dependent on all its elements. If just one primitive can be found with security flaws, the whole protocol might become insecure. Finding successful attacks against widespread cryptographic hash functions would affect a variety of popular security protocols and have unforeseeable impact on their overall security [11, 12].

In February 2005 Wang et al. presented a new attack method against the popular Secure Hash Algorithm (SHA-1). It reduces the computational attack complexity to find a collision from 2^{80} to approximately 2^{69} compression function

S. Lucks, A.-R. Sadeghi, and C. Wolf (Eds.): WEWoRC 2007, LNCS 4945, pp. 39–51, 2008.

evaluations [22] leading to the announcement that SHA-1 has been broken in theory. In 2006, it was further improved to about 2^{62} compression function calls [24]. However, still this attack is supposed to be theoretical in nature, because the necessary number of computations is very high.

For practical attacks, all theoretical results have to be mapped to an executable algorithm, which subsequently has to be launched on an appropriate architecture. Basically, there are two ways to design such architectures, namely standard and special-purpose hardware.

Generally, both FPGA and ASIC architectures require higher development costs than PC based systems. However, at high volumes special-purpose hardware is usually superior to PC clusters with respect to cost-efficiency.

The main issue of this work is whether it is possible to develop alternative hardware architectures for collision search which offer better efficiency than the aforementioned standard PC architectures. Given a certain amount of money, which hardware architecture should be invested in to gain best performance results for collision search and make practical attacks feasible?

Our solution is a highly specialized, minimal ASIC microprocessor architecture called μMD. μMD computes 32-bit words at a frequency of about 303 MHz. It supports a very small instruction set of not more than 16 instructions and, in total, can be fit on an area of just 0.022 μm^2. For collision search, μMD is connected via a 32-bit bidirectional bus to an on-chip memory and I/O module, resulting in a standalone collision search unit called μCS.

In literature, there are a few publications that deal with practical issues of collision search algorithms [3, 4]. Most of the work is dedicated to rather theoretical problems. Current implementations of MD5 collision search algorithms for PC systems are given in [8, 10, 20]. Jošćák [8] compares their performances in detail. However, other types of architectures are not considered. To our best knowledge, this is the first work that analyzes implementation requirements for collision search algorithms from an algorithmic and architectural perspective.

This work is structured as follows. Section 2 gives an introduction to the basic features of MD4-family hash functions. In contrast to this, Section 3 gives an overview over current attacking techniques on MD4-family hash functions. In Section 4, we derive from these techniques concrete design and implementation requirements for our target architecture. Subsequently, in Section 5, we give a detailed description of our final collision search architecture. In Section 6, we then develop a metric to adequately compare the performances of different hardware circuits for collision search. This metric is then applied to our final architecture, providing detailed information on its performance. We close with a short conclusion.

2   Hash Functions of the MD4-Family

A (cryptographic) hash function is an efficiently evaluable mapping h which maps arbitrary-sized messages to fixed-size hash values [7, 14]:

$$h : \{0,1\}^* \to \{0,1\}^n.$$

There are at least three features a secure hash function is expected to have; (first) preimage resistance, second preimage resistance, and collision resistance. Successful attacks on collision resistance, i.e. finding two distinct messages that map to the same hash value, are computationally much more promising, hence most attacks in the literature focus hereon.

To cope with variable input length, hash functions of the MD4-family pad the input message M such that it can be divided into fixed-size message blocks

$$M = M_1|M_2|\ldots|M_{q-1},$$

where each block M_i for $i = 1,\ldots,q-1$ is r bits long. These blocks are then successively processed by a so called compression function f.

$$f : \{0,1\}^n \times \{0,1\}^r \to \{0,1\}^n$$

To impose chaining dependencies between successive message blocks the compression function also processes the output, i.e. the so-called chaining value cv, of the preceding computation. The first chaining value cv_1 is a fixed initialization vector IV. The output of the computation of the last message block is defined to be the output of the entire hash function. Hash functions of the MD4-family differ mainly in the design of their compression function and their initialization vector.

$$cv_1 = IV$$
$$cv_{i+1} = f(cv_i, M_i),\ 1 \le i \le q - 1$$
$$h(M) = cv_q$$

Hash functions of the MD4-family are constructed in line with the Merkle-Damgård Theorem [15, 16]. This theorem delivers a useful security reduction stating that the security of a hash function can be concluded from the security of its compression function: if it is computationally infeasible to find two distinct pairs of inputs to a compression function that map to the same output (*pseudo collision*), then it is hard to find a collision of the hash function. Therefore, most attacks on hash functions of the MD4-family actually target compression functions. However, there is no general way known to exploit a single (random) pseudo collisions.

Practically, there are two ways to find useful colliding messages. They differ in the number of message blocks required to generate a collision. Single block collisions generate a collision using a single pair of distinct message blocks. Contrarily, multi-block collisions use several pairs of messages that, successively computed in a predefined order, result in a colliding output. Usually, the intermediate outputs of the compression function only differ in very few bit positions. Such situations are referred to as near collisions [1].

3 Attacks on MD4-Family Hash Functions

Generally, there are two types of attacks on MD4-family hash functions, generic and specific attacks.

Generic attacks are attacks that are applicable to all (even ideal) hash functions. Using a generic attack for finding a collision for a hash function with output size n requires computational complexity of $O(2^{\frac{n}{2}})$. This result is due to the so-called birthday attack [25]. The birthday attack basically exploits a probabilistic result that is commonly known as the birthday paradox or the birthday collision.

Specific attacks try to exploit the knowledge of the inner structure of the hash function and its inherent weaknesses. In this way, it is possible to *construct* collisions to a certain extent. Specific attacks are always dedicated to a single hash function. However, there are some general methods to develop such attacks on MD4-family hash functions.

At present, successful attacks can be divided into two phases. The first phase launches a differential attack [2] on the inner structure of the compression function [5]. Essentially, this method exploits the fact that collisions can adequately be described using differences. A collision is just a pair of messages with a difference unequal to zero that maps to a zero output difference. Differences may propagate through parts of the compression function in a predictable way. Therefore, the goal is to find conditions under which useful differences propagate with a high probability. All these identified conditions, the so-called differential path, are then mapped into a search algorithm. Assuming a random traversal over the differential path, the number of conditions reflects the search complexity of the algorithm.

The second phase of a specific attack consists of utilizing the remaining freedom of choice for the message bits. Of course, this freedom can be used to predefine parts of the input messages. However, another and very popular application is to exploit it for a *significant* acceleration of the collision search. In the literature one can find single-step modifications [23], multi-step modifications [9, 13, 19, 23], and tunneling [10]. Roughly speaking, these techniques consist of determining bits in the computation path such that, if altered in an appropriate way, they do not influence preceding, yet fulfilled conditions. Randomly choosing new message pairs and checking if all conditions up to this position are satisfied can mean high computational costs and have a high probability of failure. Instead, by exploiting these methods, it is possible to deduce new message pairs with the same characteristic based on a single pair of messages satisfying all conditions so far. Usually this deduction is computationally much cheaper and less probable to fail. Advantageously, the number of new messages increases exponentially with the number of found bits. This provides for a significant increase in efficiency.

4   Architecture Requirements

When analyzing MD4-family hash functions and their corresponding collision search algorithms, one can find several important hints how a suitable hardware architecture should be designed.

The computation process starts by randomly choosing two messages with a fixed difference, what practically rises the need for a pseudo random number generator. Then, it applies computations of the compression function (*step iterations*) to the state variables. Since hash functions of the MD4-family have been developed also with respect to PC based architectures, all step computations require similar operation sets. By implementing a few boolean, arithmetic and bit rotation operations, all requirements of the entire MD4-family can be met.

However, the collision search algorithms require additional arithmetic and flow operations. Instead of choosing new messages and recomputing all steps so far, the aforementioned acceleration techniques rather adapt new messages to fulfill conditions. Therefore, step equations have to be rearranged and solved for the message bits. For MD4-family hash functions, such rearrangements are not difficult. However, they require additional operations like the subtraction operator, which is usually not used in the original hash function specifications.

Subsequent to many computations, single data dependent bits are compared with conditions of the differential path. When satisfied, the algorithm proceeds, otherwise it returns to an earlier position in the computation path. Computationally, this requires (conditional) branches.

Hash functions of the MD4-family have been developed for fast software execution on standard PCs [18]. They operate on data units with popular processor word lengths 32-bit or 64-bit, and all their operations consist of typical processor instructions. As a consequence, the target hardware for the collision search algorithm should also work on 32-bit or 64-bit words. There are frequent operations in the hash function, like modular additions and bit rotations that do not only operate on a single pair of bits but propagate changes among neighboring bits as well. In contrast to bit-wise defined operations, like AND, OR, NOT, the actual effect of such operations heavily depends on the processor word length. When operands are divided up into subparts (e.g., chunks of 8 bits), the original impact across several bit positions may require additional processing. To guarantee compliance with the specified hash algorithm (or its collision search algorithm) these results have to be corrected in a post processing step.

Collision search algorithms can hardly be parallelized on lower hierarchical levels due to their strictly sequential structure. Useful situations are confined to operations within the step function where the evaluation of two modular additions can be computed concurrently. In almost all other cases this is not possible, since most operations also process the result of their immediate predecessor.

Besides multi-step modifications, we believe that tunneling will become a standard tool for improving collision search based on differential patterns. Unfortunately, tunneling highly parameterizes the computation path using loop constructions. This requires efficient resource reuse. In combination with frequent instruction branching, this fact renders hardware acceleration techniques like pipelining hardly useful.

For MD5 [18], there exist several efficient collision search algorithms [8, 10, 20]. Joŝĉák [8] and Stevens [21] compare their performances in more detail, where Klima's collision search (CS) approach [10] turns out to be one of the fastest.

In contrast to most of the other ones, this algorithm extensively makes use of tunneling. CS is divided into two parts, reflecting the structure of a multi collision. The first part searches for a near collision of the compression function, given the standard initialization vector. The second part generates an appropriate pseudo collision. The algorithm CS uses the standard initialization vector IV for MD5 although it can easily be altered to work with distinct initialization vectors $IV' \neq IV$. The differential path used in this algorithm is fixed and based on [13].

5   Architecture Design

5.1   Design Process

The development process can be divided into several phases. In the first step we designed the basic processor μMD using a VHDL integrated development environment.

For a verification and performance analysis, we additionally required appropriate memory devices containing the program code and constants and offering enough space for storing temporary variables. We denote the assembly of those components with μCS.

For simulation purposes, we developed tools to automatically load the ROM modules of μCS with the content of dedicated binary files. To generate these files, we had to develop a dedicated assembly language. In the next step we had to program ACS, the assembly version of CS, to gather information on the required memory space of μMD and determine some basic facts about its runtime behavior. Unfortunately, it is currently not possible to thoroughly analyze ACS long-run behavior in the simulation model for gaining average values. It is even hardly possible to observe a single collision, when the algorithm is once started.

5.2   Microprocessor Design

For the given reasons, we developed a minimal 32-bit microprocessor architecture μMD for fast collision search. It uses a very small instruction set, consisting of sixteen native commands, see Table 5 in the Appendix. In particular, this is sufficient for the execution of all algorithms of the MD4-family. Moreover, it suffices for the execution of current and (probably) future collision search algorithms, like CS. For the choice of our instruction set we compacted the results from Section 4 and used a processor reference design for μMD based on [17]. Furthermore, we designed the instruction set to maximize reuse of program code wherever possible. As a result, we also implemented a sufficiently large hardware stack and indirect load and store operations. In combination, they provide a comfortable access to parameterized subfunctions.

5.3   Collision Search Unit

μCS is our final integrated circuit for collision search. Roughly spoken, it consists of a single μMD unit, additional memory and I/O logic. To start a computation,

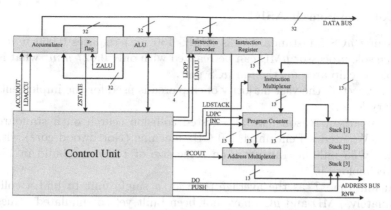

Fig. 1. Simplified architecture of collision search processor μMD

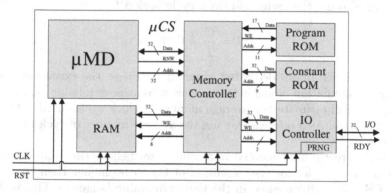

Fig. 2. Simplified schematic of collision search unit μCS

it requires an initial seed for the integrated PRNG. By carefully choosing this seed, we can guarantee that each μCS unit computes a different partition of the entire search space. When a collision is found, the corresponding message words are returned. Except for the PRNG initialization phase and the collision output sequence, there is no further I/O communication required. This decisively supports parallelization approaches making the overhead for additional control logic negligible. Figure 2 shows a simplified schematic of the collision search unit.

6   Implementation

Compared to general-purpose processors, μMD is small. μMD requires about 6k gate equivalents (GE), μCS roughly 210k GE. In an FPGA implementation, it uses about 9 percent (498 slice flip flops, 1266 4-input look-up-tables) of the slices of a low-cost Spartan3 XC3S1000 FPGA. For this device, the final clock frequency was reported with 95 MHz after synthesis.

6.1   Collision Search ASIC

Synthesizing μCS for standard cells (UMC 130 nm) requires 0.960 mm^2 chip area. The sole processor μMD can be realized with only 0.027 mm^2 what is just 2.77% of the chip area for the full μCS unit.

As we expected, the vast majority of chip area is used for the implementation of memory logic.

For comparison reasons, we also tested collision search on a standard PC processor. We used a Pentium 4, 2.0 GHz machine (Northwood core) with approximately 55 million transistors on a die size of 146 mm^2 build in 130 nm circuit technology [6].

We define time T as the average time for a single unit to find a collision. Unfortunately, μMD and μCS have not been built yet so simulated values will be used instead. For μCS this measure is computed based on the average number C of cycles required to find a collision and the corresponding frequency f. Instead of f we can also use the reciprocal clock cycle period t.

$$T = \frac{C}{f} = C \cdot t$$

When implemented in our dedicated assembly language, the execution of CS in the complete simulation model is too inefficient to achieve reliable values. In the following, we will estimate the average number of clock cycles needed to find a collision for MD5. As a reference we use the average number of cycles needed by the Pentium 4 PC.

Although directly implemented in assembly, we believe the required average number of clock cycles for the execution of CS to be higher than that of the original CS being implemented in the C programming language. This is due to two major points.

Firstly, in contrast to standard PC processors, most μMD instructions consume two clock cycles. We assume that this fact roughly doubles the number of required clock cycles compared to the Pentium 4.

Secondly, unlike Pentium 4 systems, μMD does not make use of instruction pipelining for memory access operations. This means, that store and load operations cause the ALU of μMD to halt until their completion. On μCS, these operations are not only used to load and store data but also to fetch new instructions and to control the I/O including the PRNG. We believe this fact increases the number of required clock cycles by a factor of four.

6.2   Performance Comparison

Altogether, we estimate CS executed on μMD to require roughly eight times more clock cycles than on a Pentium 4. Assuming equal production constraints (same price per chip area, see Equation 1), each of our solutions is much more effective than a comparable Pentium 4 architecture when comparing the area-time product $P = A \cdot T$.

Table 1. Processor performance - average time T to find a collision

Architecture	Cycles C	Frequency f	Period t	Time T
μCS	$480 \cdot 10^9$ cycles	102.9 MHz	9.71 ns	4660.8 s
μMD	$480 \cdot 10^9$ cycles	228.8 MHz	4.37 ns	2097.6 s
Pentium 4	$60 \cdot 10^9$ cycles	2 GHz	0.5 ns	30 s

Table 2. Processor comparison for area-time product P

Architecture	Time T	Area A	$P = A \cdot T$
μMD	2097.6 s	0.027 mm^2	55.9
Pentium 4	30 s	146 mm^2	4380

Table 2 compares the performance characteristics of μMD with those of a standard Pentium 4 processor. It is obvious, that μMD has better characteristics than the Pentium 4 processor. The performance of μMD for collision search is about 62 times better than the Pentium's. When we use an off-the-shelf standard PC for parallelization, we have to consider the costs for a motherboard with a network card that supports Preboot Execution Environment (PXE) (€ 80), fan (€ 12), RAM (€ 25), power supply (€ 25), additional equipment for the network infrastructure (network cables, switches), and a control server. Using these standard PCs, the processors are connected to each other by standard network equipment. Altogether, we believe the costs to be approximately € 200, whereas we assume the Pentium 4 2.0 GHz to have a price of roughly € 50 leading to an overall parallelization overhead O_p of 300%. From a Pentium 4 processor, we estimate the price per chip Q_A area to be

$$Q_A = \frac{50 \text{ €}}{146 \text{ mm}^2} = 0.343 \frac{\text{€}}{\text{mm}^2} \ . \tag{1}$$

The parallelization of μCS however, requires only few additional logic per unit. In combination with low-throughput bus connection of μCS units, this provides an optimal scaling solution without noticeable additional costs. For our solution, we assume the area overhead to be almost negligible (less than 5%).

Based on these considerations, we believe that our *full* collision search solution is noticeably more effective for collision search than parallelized Pentium 4 processors.

Table 3 presents our estimates for the design where Q_s reflects the price for a single standalone unit of the corresponding architecture. In contrast, Q_p is the average price for a single unit after parallelization.

Obviously, for finding a single MD5 collision per second one has to spend € 6000 in (30) parallelized standard PCs (see Figure 3). Assuming similar constraints for manufacturing ASICs and excluding any NRE costs, the asset cost for the same performance invested in parallelized μCS units is about € 1600. So, it is almost four times more cost-efficient than the Pentium 4 standard PC architecture (see Table 4).

Table 3. Comparison of Pentium 4 and μCS architectures

Metric	Pentium 4	μCS
Costs per area Q_A	$0.343 \frac{\text{€}}{mm^2}$	$0.343 \frac{\text{€}}{mm^2}$
Chip area A	$146 \ mm^2$	$0.960 \ mm^2$
Chip cost Q_s	50 €	0.33 €
Overhead O_p	300 %	5 %
System cost Q_p	200 €	0.35 €
AT-Product P	4380	4472.6
Time T per unit	30 s	4660.8 s
Costs R for a collision per second	6000 €	1608.40 €

Table 4. Performance ratio of μCS compared to Pentium 4

Architecture	Costs R for a collision per second	Ratio R/R_{P4}
Pentium 4 PC	6000 €	100 %
μCS	1608.4 €	26.8 %

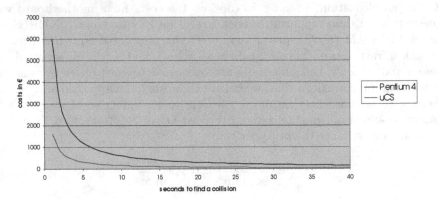

Fig. 3. Costs for equipment to find a MD5 collision per time

6.3   Estimates for SHA-1

We believe that a comparable implementation of a SHA-1 collision search algorithm in dedicated and parallelized collision search units has even better performance characteristics. This is mainly due to the fact, that SHA-1 needs much less constants than MD5, thus radically reducing the costs for (constant) ROMs. We assume that a collision search algorithm for SHA-1 can be programmed similarly compact. Although SHA-1 spans more steps, what surely the size of the corresponding collision search algorithm, the program code will not considerably reflect this. This can be concluded from the unique implementation of subroutines which can be called on demand using only few additional instructions.

The average number of required clock cycles to find a collision is primarily dependent on the available theoretical results. Currently, attacks on SHA-1 have a complexity of about 2^{62} compression function evaluations. For practical attacks, we believe this number to be still very large.

Given € 1 million, an attacker can buy enough standard PC equipment to find a MD5 collision within 6 ms on average based on the assumptions above. Invested in our collision search unit it would take only 1.61 ms. In [8], finding MD5 collisions based on CS has been reported to have a complexity of about $7.7 \cdot 2^{30}$ MD5 step operations. The current bound for SHA-1 collisions is 2^{62} compression function evaluations, while each such evaluation is composed of 80 step function evaluations. Assuming an actual complexity bound of 2^{70} step operations and a similar execution time for a single step operation in MD5 and SHA-1, finding collisions for SHA-1 takes about 2^{37} times more than for MD5. We can also conclude how long it would take to find a single SHA-1 collision with equipment for € 1 million. Invested in standard PCs, it would take 6 ms$\cdot 2^{37} \approx 26$ years. Using our collision search units, this time would be only 1.61 ms$\cdot 2^{37}$ s ≈ 7 years. Assuming Moore's law to hold for the next years, a successful attack for € 1 million in one year using a parallel μCS architecture should be possible in 2012 (less than next five years).

7   Conclusion

In this work we analyzed the hardware requirements of current and future collision search algorithms for hash functions of the MD4-family. We used our results to develop an appropriate hardware platform.

The heart of our design is a very small microprocessor μMD with only sixteen instructions. At the same time, it provides very effective means to support program code reuse, what greatly helps to keep the size of our overall collision search unit μCS small.

In the context of MD4-family hash functions, μMD is general-purpose, meaning that it is appropriate for the execution of all (32-bit) MD4-family hash functions and also of all corresponding current and future collision search algorithms.

In contrast to standard PCs, the final collision search unit needs only very little additional logic. This reduces its price and greatly eases parallelization approaches.

We believe that our design approach is much better suited for collision search than standard PCs. When money is spent on collision search, our design, massively parallelized, is nearly four times more cost-efficient than parallelized P4 standard PCs. Since this performance factor with respect to general-purpose processors is not too significant, further optimizations like an improved instruction fetch unit might lead to a better performance. These optimizations might be necessary to justify the additional efforts (like NRE and manufacturing costs) when favoring a special-purpose hash collision unit over standard off-the-shelf processors.

References

1. Biham, E., Chen, R.: Near-Collisions of SHA-0. In: Franklin, M. (ed.) CRYPTO 2004. LNCS, vol. 3152. pp. 290–305. Springer, Heidelberg (2004)
2. Biham, E., Shamir, A.: Differential Cryptanalysis of DES-like Cryptosystems. In: Menezes, A., Vanstone, S.A. (eds.) CRYPTO 1990. LNCS, vol. 537, pp. 2–21. Springer, Heidelberg (1990)
3. De Cannière, C., Mendel, F., Rechberger, C.: On the Full Cost of Collision Search for SHA-1. Presentation at ECRYPT Hash Workshop 2007 (May 2007)
4. De Cannière, C., Rechberger, C.: Finding SHA-1 Characteristics: General Results and Applications. In: Lai, X., Chen, K. (eds.) ASIACRYPT 2006. LNCS, vol. 4284. pp. 1–20. Springer, Heidelberg (2006)
5. Chabaud, F., Joux, A.: Differential Collisions in SHA-0. In: Krawczyk, H. (ed.) CRYPTO 1998. LNCS, vol. 1462. pp. 56–71. Springer, Heidelberg (1998)
6. Intel Corporation. Intel Pentium 4 Processor Specification Update (May 2007), http://www.intel.com
7. M. Daum. Cryptanalysis of Hash Functions of the MD4-Family. PhD thesis, Ruhr-Universität Bochum (2005), http://www.cits.rub.de/MD5Collisions/
8. Jošćák, D.: Finding Collisions in Cryptographic Hash Functions. Master's thesis, Univerzita Karlova v Praze (2006), http://cryptography.hyperlink.cz/2006/diplomka.pdf
9. Klima, V.: Project Homepage (2006), http://cryptography.hyperlink.cz/MD5_collisions.html
10. Klima, V.: Tunnels in Hash Functions: MD5 Collisions Within a Minute. Cryptology ePrint Archive, Report 2006/105 (2006), http://eprint.iacr.org/
11. Lenstra, A., de Weger, B.: On the possibility of constructing meaningful hash collisions for public keys. In: Boyd, C., González Nieto, J.M. (eds.) ACISP 2005. LNCS, vol. 3574. Springer, Heidelberg (2005)
12. Lenstra, A., Wang, X., de Weger, B.: Colliding X.509 Certificates (2005), http://eprint.iacr.org/
13. Liang, J., Lai, X.: Improved Collision Attack on Hash Function MD5. Cryptology ePrint Archive, Report 2005/425 (November 2005), http://eprint.iacr.org/
14. Menezes, A.J., van Oorschot, P.C., Vanstone, S.A.: Handbook of Applied Cryptography. CRC Press, Boca Raton (1997)
15. Merkle, R.: One Way Hash Functions and DES. In: Brassard, G. (ed.) CRYPTO 1990. LNCS, vol. 435. Springer, Heidelberg (1990)
16. Damgård, I.: A Design Principle for Hash Functions. In: Brassard, G. (ed.) CRYPTO 1990. LNCS, vol. 435. Springer, Heidelberg (1990)
17. Reichardt, J., Schwarz, B.: VHDL-Synthese, 3rd edn. Oldenbourg (2003)
18. Rivest, R.: The MD5 Message-Digest Algorithm, Request for Comments (RFC) 1321 (1992), http://www.ietf.org/rfc.html
19. Sasaki, Y., Naito, Y., Kunihiro, N., Ohta, K.: Improved Collision Attack on MD5. Cryptology ePrint Archive, Report 2005/400 (November 2005), http://eprint.iacr.org/
20. Stevens, M.: Fast Collision Attack on MD5. Cryptology ePrint Archive, Report 2006/104 (2006), http://eprint.iacr.org/
21. Stevens, M.: On Collisions for MD5. Master's thesis, Eindhoven University of Technology, Department of Mathematics and Computing Science (June 2007)
22. Wang, X., Yin, Y.L., Yu, X.: Finding Collisions in the Full SHA-1. In: Shoup, V. (ed.) CRYPTO 2005. LNCS, vol. 3621, pp. 17–36. Springer, Heidelberg (2005)

23. Wang, X., Yu, X.: How to Break MD5 and other Hash Functions. In: Cramer, R. (ed.) EUROCRYPT 2005. LNCS, vol. 3494. pp. 19–35. Springer, Heidelberg (2005)
24. Wang, X.: Cryptanalysis on hash functions. Presentation at Information-Technology Promotion Agency (IPA), Japan (October 2006),
    http://www.ipa.go.jp/security/event/2006/crypt-forum/pdf/Lecture_4.pdf
25. Yuval, G.: How to Swindle Rabin. Cryptologia 3(3), 187–189 (1979)

Appendix: Instruction Set for μMD

Table 5. Instruction set of μMD processor

Opcode	Value	Z-flag	Cycles	Description
RL	0x0000	not altered	2	Rotate A's bits to the left. The rotation width is found at the specified memory address.
STA	0x0001	not altered	2	Store A to absolute memory or IO address
STI	0x0010	not altered	3	Store A to indirect memory or IO address
LDI	0x0011	not altered	3	Load A from indirect memory or IO address
LDA	0x0100	not altered	2	Load A from absolute memory or IO address
ADD	0x0101	possibly altered	2	Add $(\bmod 2^{32})$ specified memory word to A
SUB	0x0110	possibly altered	2	Subtract $(\bmod 2^{32})$ the specified memory word from A
OR	0x0111	possibly altered	2	Compute logical OR of A and specified memory word
AND	0x1000	possibly altered	2	Compute logical AND of A and specified memory word
XOR	0x1001	possibly altered	2	Compute logical XOR of A and specified memory word
JMP	0x1010	not altered	2	Jump to specified address
JE	0x1011	not altered	2	Jump to specified address if z-flag is set to '0'
JNE	0x1100	not altered	2	Jump to specified address if z-flag is not set to '0'
CALL	0x1101	not altered	2	Push incremented program counter onto the stack and jump to specified address
RET	0x1110	not altered	3	Jump to address stored in top of stack. Pop top of stack
NOT	0x1111	possibly altered	1	Compute logical NOT of A

Cryptography Based on Quadratic Forms: Complexity Considerations

Rupert J. Hartung

Johann Wolfgang Goethe Universität Frankfurt a. M.
Postfach 11 19 32; Fach 238
60054 Frankfurt a. M., Germany
hartung@mi.informatik.uni-frankfurt.de
http://www.mi.informatik.uni-frankfurt.de

Abstract. We study the computational problem **Trafo** of finding an integral equivalence transform between two given quadratic forms. This is motivated by a recent identification scheme based on this problem [10]. We prove that for indefinite forms over $\mathbb{Z}$, its hardness is concentrated in dimensions 3 and 4. Moreover, over the field of rational numbers the complexity of **Trafo** is closely related to that of factoring. However, for definite forms over $\mathbb{Z}$, as well as for forms over finite fields, the transformation problem is solvable in polynomial time.

1 Introduction

Lattice-based cryptography has been a vivid field in cryptologic research ever since its proposal (see [2], [12], [13], [11], [9]). Lattice cryptosystems are based on variants of either the Shortest Vector Problem (SVP) of finding a shortest non-zero vector in a given lattice, or the Closest Vector Problem (CVP) of finding the closest vector in a given lattice relative to a given point in space. A main advantage of this family of cryptosystems is considered to be the fact that SVP and CVP have been proven to be NP-complete (partially under randomized reductions, see [16], [17]). This is taken as a hint that this type of primitives may still be secure in the (still hypothetical) age of quantum computers because cryptographic schemes based on NP-hard problems seem to withstand attacks by quantum computers (see [3]).

However, the NP-hardness proofs fail for these problems in bounded dimension. Accordingly, lattice cryptography intrinsically requires lattices of high dimension. This leads to large cryptographic keys and slow schemes.

Lattices correspond to definite quadratic forms via their Gram matrices. Thus turning towards indefinite forms may be the right idea to overcome the problems of lattice cryptography. Indeed, in [10], an identification scheme has been proposed based on the transformation problem on indefinite quadratic forms of dimension 3 over $\mathbb{Z}$. It was shown there that under a reasonable number-theoretic assumption, this problem is NP-hard under randomized reductions if restricted to small transforms. This strengthens the analogy to lattice cryptography. By contrast, the identification scheme from [10] uses only few arithmetic operations per round, and by the argumentation there keys of 200 or 300 bits in size should be sufficient for practical security of the scheme.

The aim of this article now is to abstract from this concrete scheme and the specific suggestions on parameters and keys. We consider the underlying problem, the *transformation problem*, and study its complexity for various families of quadratic forms over various rings R. It will turn out that the choice of indefinite forms over $R = \mathbb{Z}$ of dimension $n = 3$ taken in [10] is in some sense 'optimal' (among forms over $\mathbb{Z}$, $\mathbb{Q}$, and finite fields). This may shed some light on how to design other cryptographic schemes based on quadratic forms.

Outline. This article is organized as follows: In Sect. 2, we introduce quadratic forms together with the computational problem **Trafo**; subsequently, we discuss the cryptographic application. Our main results are contained in Sect. 3 and Sect. 4. In Sect. 3, we study **Trafo** over fields. We show that over prime fields, **Trafo** is tractable in random polynomial time; by contrast, over the rationals $\mathbb{Q}$, the transformation problem is closely linked to factoring. Then, in Sect. 4, we show that **Trafo**$^{\mathbb{Z}}$ does not become any easier if restricted to dimensions 3 and 4. Finally, we mention the hardness result on a variant of **Trafo**.

This paper makes essential use of the arithmetic theory of quadratic forms. A comprehensive account of this topic can be found in the textbook by Cassels [5], which we will frequently refer to.

2   Preliminaries

2.1   Quadratic Forms

A *quadratic form* (or simply *form*) is a homogeneous polynomial of degree 2 over some unique factorization domain R, which we assume not to be of characteristic 2; thus $f = x^t A x$ with $A \in \frac{1}{2} R^{n \times n}$ symmetric, with integral main diagonal, and an indeterminate vector $x = (x_1, \ldots, x_n)^t$. If A is a diagonal matrix, then we use the abbreviation $f = \langle a_1, \ldots, a_n \rangle := \sum_{i=1}^{n} a_i x_i^2$. If f, g are forms, their orthogonal sum $f \perp g$ is the form $f(x) + g(y)$, where x and y are disjoint sets of variables. The number of variables n is called the *dimension*, and

$$\det f := \det A$$

the *determinant* of the quadratic form f. For $T \in R^{n \times n}$ let fT denote the form $x^t T^t A T x$. Two forms f, g are called *equivalent* (or *R-equivalent*) over R if there is a matrix $T \in \mathrm{GL}_n R$ such that $g = fT$; denoted by $f \sim_R fT$. The equivalence class of f is simply called the *class* of f. Note that $\mathrm{GL}_n \mathbb{Z} = \{ S \in \mathbb{Z}^{n \times n} \mid \det S = \pm 1 \}$ and $\mathrm{GL}_n K = \{ S \in K^{n \times n} \mid \det S \neq 0 \}$ for a field K.

A quadratic form f is said to *represent* $m \in R$ if there is $v \in R^n \setminus \{0\}$ such that $f(v) = m$. It is said to represent m *primitively* if v can be chosen primitive, *i.e.* $\gcd(v_1, \ldots, v_n) = 1$.

A quadratic form f (over $\mathbb{Q}$ or $\mathbb{Z}$) is called *indefinite* if it represents both positive and negative numbers, and *definite* if it represents either only positive or only negative numbers. Lattices are closely related to positive definite forms. A quadratic form is called *isotropic* if it represents 0, and anisotropic otherwise.

2.2   The Transformation Problem

We consider the following computational problem. In the crypto scheme of Sect. 2.3, it will appear as the problem of extracting the secret key from the public key.

> **Trafo$^R(\mathcal{P})$ The transformation problem over R**
> *PARAMETERS:* Set $\mathcal{P}$ of properties of quadratic forms, domain R.
> *INPUT:* quadratic forms f, g satifying all properties from $\mathcal{P}$ and $f \sim_R g$.
> *OUTPUT:* $S \in \mathrm{GL}_n R$ such that $g = f\,S$ (where n is the dimension of f).

The parameter set $\mathcal{P}$ allows us to restrict to forms satisfying certain properties, *e.g.* indefiniteness, or $\dim f = 3$. For each choice of $\mathcal{P}$, we regard **Trafo**$^R(\mathcal{P})$ as a computational problem on its own.

Note that we do not insist that a specific S should be constructed. In the identification scheme below, any matrix which is a solution to the transformation problem would enable a deceiver to pass (*i.e.* break) the scheme.

The transformation problem on definite forms. Note that for fixed dimension, the transformation problem for definite forms can be solved in polynomial-time (although the complexity seems to be devastating as the dimension increases). We shall use this fact later on. The key ideas are due to Plesken and Pohst [19], see also the discussion in [20].

As a subroutine, they use a method to enumerate short vectors of a lattice, in particular, to solve the SVP (see [15]). It is a vivid line of research to improve SVP algorithms and their analysis [24], and thus to make the transformation procedure tractable in slightly higher dimensions. However, as we are interested in problems in fixed low dimensions, we view this problem as settled.

2.3   Cryptographic Application

In [10], an identification scheme has been proposed according to the following outline:

Public key: Equivalent forms f_0, f_1,
Secret key: $S \in \mathrm{GL}_n R$ such that $f_0 S = f_1$.

1. Prover $\mathcal{P}$ picks $T \in R^{n \times n}$ (according to some distribution); computes $g := f_0 T$, and sends g,
2. Verifier $\mathcal{V}$ sends a random one-bit challenge $b \in_R \{0, 1\}$,
3. $\mathcal{P}$ sends $Q := S^{-b}T$, and $\mathcal{V}$ checks that $f_b Q = g$.

As noted beforehand, this is the generic outline of the scheme. The concrete proposal in [10] uses the concept of LLL-reduction for indefinite quadratic forms (see [23] and [14]) as a key ingredient to smoothen the employed probability distributions. According to [10], the scheme then is statistically zero-knowledge for $R = \mathbb{Z}$, $n = 3$ under some additional hypothesis on the distribution in question.

Moreover, the generic scheme as presesented here is a proof of knowledge in general, in the sense that a fraudulent prover $\widetilde{\mathcal{P}}$ which passes the scheme with the same commitment g on both challenges $b = 0, 1$ can compute an equivalent secret key. Namely, if $\tilde{\mathcal{P}}$ replies with matrices Q_0, Q_1, satisfying

$$f_0 \, Q_0 = g \qquad \text{and} \qquad f_1 \, Q_1 = g,$$

then the matrix $S' := Q_0 Q_1^{-1}$ satisfies

$$f_0 \, S' = f_1,$$

which is exactly the equation characterizing the secret key.

It should be noted that the fraudulent prover's trivial probability of success amounts to as much as $\frac{1}{2}$, simply by guessing the challenge b in advance. Therefore, the protocol has to be repeated polynomially many times to make his chance negligible.

It is possible that there are other suitable choices of parameters in this generic protocol than those of [10]. For instance, a different random distribution on integer matrices might allow for a simple proof of the zero-knowledge property.

In this paper, we shall ignore issues about random distributions and concentrate on how the complexity of the underlying computational problem changes with the use of different rings and families of quadratic forms. Since the protocol is a proof of knowledge, the study of the underlying problem means determining the security agains malicious provers. We will come to the conclusion that the choice $R = \mathbb{Z}$, $n = 3$ is in some sense the "right" choice: Choosing a different dimension than 3 or 4 does not add to the complexity of the problem, whereas using forms over the rationals $\mathbb{Q}$, or over finite fields, is likely to decrease complexity.

We shortly note that there may be other alternatives not considered here. For instance, indefinite quadratic forms over the ring of integers of an algebraic number field may be a good candidate as they feature many similarities with those over $\mathbb{Z}$, see [18].

3   Choice of the Base Ring

For forms over fields we need the well-known

Lemma 1 (Completion of the square). *Let K be a field of characteristic $\neq$ 2. Then every quadratic form over K is equivalent to a diagonal form.- Moreover, for fixed dimension, the transformation implied can be computed in polynomial time.*

Proof. See [5, ch. 2, lm. 1.4]. □

Though polynomial time in theory, diagonalization over $\mathbb{Q}$ may yield coefficients with prohibitively large enumerators and denominators in practice, see [23, introd.]. This is due to the fact that the number of steps performed may be exponential in n, where n is the dimension of the form. Note that Lemma 1 only makes a statement for fixed n.

3.1   Finite Fields

By **Trafo**$^{\mathbb{F}}$ with the property set omitted, we mean the transformation problem on *all* regular quadratic forms over $\mathbb{F}$ without restrictions.

Theorem 1. *Let p be an odd prime and let $\mathbb{F}$ be the finite field with p elements. Then* **Trafo**$^{\mathbb{F}}$ *is solvable in random polynomial time.*

Proof. Let $\mathbb{F}$-equivalent forms f, g over $\mathbb{F}$ be given. Use Lemma 1 to transform each of them into diagonal shape, say

$$f\,T_1 = \langle a_1, \ldots, a_n \rangle \quad \text{and} \quad g\,T_2 = \langle b_1, \ldots, b_n \rangle$$

with $T_i \in \mathrm{GL}_n\mathbb{F}$. Then determine which of the a_i, b_i are squares in $\mathbb{F}$. This can be done in polynomial time by computing Jacobi symbols. Build a permutation matrix Π_i for each form such that $f\,T_1\Pi_1 = \langle s_1, \ldots, s_k, q_{k+1}, \ldots, q_n \rangle$ and $g\,T_2\Pi_2 = \langle s_1', \ldots, s_\ell', q_{\ell+1}', \ldots, q_n' \rangle$ such that all q_i, q_i' are squares and all s_i, s_i' are non-squares. Without loss of generality, we may assume that $k \geq \ell$. Let $s := s_1$ (if $k \neq 0$). Then $s_i^{-1}s$, $(s_i')^{-1}s$ are square for all i. By [8, sec. 2.3.2], we can compute square roots

$$r_i^2 = q_i, \qquad (r_i')^2 = q_i', \qquad t_i^2 = s_i^{-1}s, \qquad \text{and} \qquad (t_i')^2 = (s_i')^{-1}s$$

in random polynomial time for all i for which the respective right hand side is defined. Then

$$f' := f\,T_1\Pi_1 S_1 = \langle \underbrace{s, \ldots, s}_{k \text{ times}}, \underbrace{1, \ldots, 1}_{n-k \text{ times}} \rangle, \quad \text{and}$$

$$g' := g\,T_2\Pi_2 S_2 = \langle \underbrace{s, \ldots, s}_{\ell \text{ times}}, \underbrace{1, \ldots, 1}_{n-\ell \text{ times}} \rangle,$$

where S_1, S_2 are diagonal matrices with diagonal entries $t_1, \ldots, t_k, r_{k+1}, \ldots, r_n$ and $t_1', \ldots, t_\ell', r_{\ell+1}', \ldots, r_n'$, respectively. As f' and g' are $\mathbb{F}$-equivalent, their determinant may only differ by a square in $\mathbb{F}$. Hence $k - \ell$ is even. Therefore it suffices for the completion of the algorithm to construct a transformation of the form $\langle s, s \rangle$ into the form $\langle 1, 1 \rangle$.

Note that there are $x, y \in \mathbb{F}$ satisfying $sx^2 + sy^2 = 1$ by [5, ch. 2, lm. 2.2]. Such a solution can be found efficiently *e.g.* using [1]. Then find x', y' such that

$$\det \begin{pmatrix} x & x' \\ y & y' \end{pmatrix} = s^{-1}$$

(*e.g.* by choosing x' uniformly at random and solving for y', until success). Then the form

$$h := \langle s, s \rangle \begin{pmatrix} x & x' \\ y & y' \end{pmatrix}$$

has the right first coefficient and the right determinant. Therefore, another application of Lemma 1 transforms it into the form $\langle 1, 1 \rangle$. This completes the description of the algorithm. As argued in the single steps, it runs in random polynomial time. $\qquad \square$

3.2   The Rational Number Field

For the field of rational numbers, by contrast, we observe that solving the transformation problem is essentially equivalent to factoring integers. This is made precise by Theorem 3.

The classical theory of quadratic forms provides us with quite strong tools to decide equivalence of rational quadratic forms. We review some important facts from [5, ch. 3,4,6].

By Lemma 1, every form f over $\mathbb{Q}$ is equivalent to a form $\langle a_1, \ldots, a_n \rangle$. For every prime number p, the *Hasse-Minkowski invariant* $c_p(f)$ is an invariant of the class of f; it is defined by

$$c_p(f) = \prod_{1 \leq i < j \leq n} \left(\frac{a_i, a_j}{p} \right),$$

where $\left(\frac{a, b}{p} \right)$ is the *norm residue symbol* modulo p: It satisfies $\left(\frac{a, b}{p} \right) = 1$ if and only if

$$a = x^2 - by^2$$

is solvable for x, y in the field $\mathbb{Q}_p$ of p-adic numbers, and $\left(\frac{a, b}{p} \right) = -1$ otherwise [5, sec. 3.2]. We only need the following rules to compute it: The symbol is bilinear, *i.e.*

$$\left(\frac{a, bc}{p} \right) = \left(\frac{a, b}{p} \right) \left(\frac{a, c}{p} \right), \qquad \text{and} \tag{1}$$

$$\left(\frac{1, a}{p} \right) = 1 \tag{2}$$

for all $a, b, c \in \mathbb{Q} \backslash \{0\}$; moreover, if p is odd, $a, b \in \mathbb{Z}$ and $p \nmid a, b$, then

$$\left(\frac{a, b}{p} \right) = 1, \qquad \text{and} \tag{3}$$

$$\left(\frac{a, p}{p} \right) = \left(\frac{a}{p} \right), \tag{4}$$

where

$$\left(\frac{a}{p} \right) = \begin{cases} 1 & \text{if } a \text{ is a square mod } p \text{ (and } p \nmid a), \\ 0 & \text{if } p | a, \\ -1 & \text{else} \end{cases}$$

is the Legendre symbol modulo p.

Now consider the field $\mathbb{R}$ of real numbers. A form $\langle a_1, \ldots, a_n \rangle$ can be easily transformed into the form $\langle \text{sign}(a_1), \ldots, sign(a_n) \rangle$ over $\mathbb{R}$ because every positive real number has a square root in $\mathbb{R}$.

Let f, g be rational quadratic forms of the same dimension and the same integral determinant. If they are equivalent over $\mathbb{Q}$, they obviously have to be equivalent over $\mathbb{R}$. The converse holds in the following sense:

Theorem 2 (Hasse Principle). *Let f, g be quadratic forms of dimension n and determinant d with integral coefficients. Then f and g are $\mathbb{Q}$-equivalent if and only if*

(i) they are $\mathbb{R}$-equivalent, and
(ii) $c_p(f) = c_p(g)$ for all primes $p \mid d$.

See [5, ch. 6] for a proof.   □

Let $p, q \equiv 1 \mod 4$ be distinct primes, and let $N := pq$. Recall that then the equation

$$x^2 \equiv -1 \mod N$$

is solvable (for $x \in \mathbb{Z}$). The integer factorization problem is not likely to become significantly easier if restricted to such numbers; compare it to the well-known hardness hypothesis for Blum integers [4]. Denote by **Sqrt** the problem of computing a square root of -1 modulo a given $N = pq$, $p, q \equiv 1 \mod 4$.

Denote by $\preccurlyeq$ polynomial-time reducibility and by $\preccurlyeq_r$ random polynomial-time reducibility of computational problems. We write **Trafo**$_n^{\mathbb{Q}}$ for **Trafo**$^{\mathbb{Q}}(\mathcal{P}_n)$, where $\mathcal{P}_n$ stands for the property $\dim f = n$.

Theorem 3. *Let $\mathcal{P}$ consist of the properties of being indefinite anisotropic. Then*

*(a) **Sqrt** $\preccurlyeq_r$ **Trafo**$^{\mathbb{Q}}(\mathcal{P})$. One oracle call suffices.*
*(b) Denote by **Fact** the problem of factoring integers and let $n \in \mathbb{N}$. Then*

$$\textbf{Trafo}_n^{\mathbb{Q}} \preccurlyeq \textbf{Fact}.$$

*For an instance (f, g) of **Trafo**$_n^{\mathbb{Q}}$, it suffices to call the oracle once to factor $(\det f)(\det g)$.*

As there is no modular square root algorithm known which is essentially faster than factoring the modulus, we may informally state that the transformation problem is 'almost' equivalent to factoring.

The problem **Sqrt** should not be confused with the problem of computing arbitrary square roots modulo N, which is known to be probabilistically equivalent to factoring N. However, to find a root of -1 only, there might theoretically be a different method.

Proof. (a) Let N be an instance of **Sqrt**. Define

$$f := \langle 1, -N \rangle, \qquad g := \langle -1, N \rangle.$$

Obviously, these are indefinite, and they are anisotropic by [5, ch. 4, lm. 2.4]. We claim that f and g are $\mathbb{Q}$-equivalent. Obviously, f and g are equivalent over the reals, as both are $\mathbb{R}$-equivalent to the form $\langle 1, -1 \rangle$. Moreover, we can compute Hasse-Minkowski invariants as follows: Let $N = pq$. Then

$$c_p(f) = \left(\frac{1, -N}{p} \right) = 1$$

by (2), and

$$c_p(g) = \left(\frac{-1, N}{p}\right) = \left(\frac{-1, p}{p}\right) \underbrace{\left(\frac{-1, q}{p}\right)}_{=1} = \left(\frac{-1}{p}\right) = 1.$$

Here the first equality follows from (1), the next one is due to (3) and (4) and the last equality sign holds because -1 is a square modulo N. An analogous computation works for q instead of p. Hence by the Hasse principle Theorem 2 $f \sim_{\mathbb{Q}} g$.

Now let $S=(s_{ij}) \in \mathrm{GL}_n\mathbb{Q}$ satisfy $f\, S=g$. Then $S^t \begin{pmatrix} 1 & \\ & -N \end{pmatrix} S = \begin{pmatrix} -1 & \\ & N \end{pmatrix}$, whence $s_{11}^2 - s_{21}^2 N = -1$. Compute minimal $k \in \mathbb{N}$ such that $\sigma_{11} := ks_{11}$, $\sigma_{21} := ks_{21}$ are integers (via the Euclidean Algorithm). Write $k = N^\ell k_0$ with $N \nmid k_0$. Then the equation

$$\sigma_{11}^2 - \sigma_{21}^2 N = -k_0^2 N^{2\ell} \tag{5}$$

holds in $\mathbb{Z}$. We claim that we can assume $\ell = 0$ without loss: Indeed, otherwise (5) implies $\sigma_{11} = \sigma_{11}' N$ for some $\sigma_{11}' \in \mathbb{Z}$. Hence

$$(\sigma_{11}')^2 N - \sigma_{21}^2 = -k_0^2 N^{2\ell-1}.$$

But now it follows that $\sigma_{21} = \sigma_{21}' N$ with $\sigma_{21}' \in \mathbb{Z}$. Thus

$$(\sigma_{11}')^2 - (\sigma_{21}')^2 N = -k_0^2 N^{2(\ell-1)},$$

analogously to (5). Inductively we can achieve $\ell = 0$.

Now (5) with $\ell = 0$ implies

$$\sigma_{11}^2 \equiv -k_0^2 \mod N. \tag{6}$$

If $\gamma := \gcd(k_0, N) \neq 1$, then γ is p or q, and the factorization of N allows to compute square roots of -1 modulo p and q, and combine them by means of the Chinese Remainder Theorem.

Otherwise, $\gcd(k_0, N) = 1$, and we can compute $\bar{k} \in \mathbb{Z}$ such that $\bar{k} k_0 \equiv 1 \mod N$. Then (6) implies that $(\sigma_{11}\bar{k})^2 \equiv -1 \mod N$.

(b) Let (f, g) be an instance of $\mathbf{Trafo}^{\mathbb{Q}}$, and let $n := \dim f$. Then $\phi := f \perp (-g)$ is isotropic (recall that $f \perp (-g)$ is the form $f(x) - g(y)$, where $x = (x_1, \ldots, x_n)^t$, $y = (y_1, \ldots, y_n)^t$ are disjoint vectors of variables).

Retrieve the factorization of $\det \phi = (-1)^n (\det f)(\det g)$ from the oracle. Then an algorithm by Simon ([22], see also [23]) constructs an isotropic vector $(0, 0)^t \neq (v_1, v_2)^t \in \mathbb{Q}^{2n}$ for ϕ in deterministic polynomial time.

If $f(v_1) = 0$ then also $g(v_2) = 0$. At least one of the $v_i \neq 0$, hence both f and g are isotropic as $f \sim g$. If $v_1 \neq 0 \neq v_2$, then these are isotropic vectors for f, g; if without loss $v_2 = 0$, then the factorization of $\det g$ can be employed again to construct a primitive isotropic vector v_2' for g. Now if v_1, v_2 are isotropic vectors for f, g, it is easy and well-known that one can find matrices $H_i \in \mathrm{GL}_n \mathbb{Q}$ with $f H_1 = h_0 \perp f_1$ and $g H_2 = h_0 \perp g_1$ for some $(n-2)$-ary forms f_1, g_1, where h_0 is the "hyperbolic plane" $\left(\begin{smallmatrix} 0 & 1 \\ 1 & 0 \end{smallmatrix} \right)$, see [5, ch. 2, lm. 2.1 and its cor. 1]. By Witt's lemma ([5, ch.2, cor. 1 of thm. 4.1]), $f_1 \sim_\mathbb{Q} g_1$. We recursively call the procedure outlined here, yielding eventually a transformation $f_1 S = g_1$. Then return

$$T := H_1 \begin{pmatrix} 1 & & \\ & 1 & \\ & & S \end{pmatrix} H_2^{-1}$$

since then $f T = g$.

If, however, $a := f(v_1) \neq 0$, then also $g(v_2) = a$. Extend the vectors v_1 and v_2, respectively, to bases of $\mathbb{Q}^n$, so that we obtain matrices $U_i \in \mathrm{GL}_n \mathbb{Q}$ with

$$f U_1 = \langle a \rangle \perp f_2 \qquad \text{and} \qquad g U_2 = \langle a \rangle \perp g_2$$

for $(n-1)$-ary forms f_2, g_2, and if the recursion produces $f_2 S = g_2$, then output $T := U_1 \left(\begin{smallmatrix} 1 & 0 \\ 0 & S \end{smallmatrix} \right) U_2^{-1}$.

Finally, this recursion will be called at most n times so that we have established a polynomial-time algorithm. $\square$

4   Concentration in Dimensions 3, 4

Let us now turn to the transformation problem over $\mathbb{Z}$. In the last section we have seen that the factorization of the determinant is closely related to the transformation problem over $\mathbb{Q}$. By a similar argument as there it can be shown that **Trafo** is at least as hard as factoring. However, as all existent algorithms for **Trafo** involve exhaustive search on a set of exponential size, the factorization of the determinant seems not to help much in finding transformations, and we want to separate this moderate obstacle from the actual core of the problem.

Thus denote by **FTrafo** instead of **Trafo** the problems where the factorization of the determinants of all given forms is included in the input. From now on, we drop the superscript R, as we restrict ourselves to $R = \mathbb{Z}$.

Theorem 4. *Denote by $\mathcal{F}_n(d)$ the properties $\det f = d$ and $\dim f = n$ for a quadratic form f.*

Let $n \geq 5$. let $d \in \mathbb{Z}$ be odd and squarefree and let the factorization of d be given. Then

$$\mathbf{FTrafo}(\mathcal{F}_n(d)) \preceq \mathbf{FTrafo}(\mathcal{F}_{n-2}(d))$$

This has the following striking consequence.

Corollary 1. *Let $\mathcal{F}$ denote the properties "$\det f$ is odd and squarefree" and "$\dim f \geq 3$" for a quadratic form f, and $\mathcal{F}_{3,4}$ for additionally "$\dim f \in \{3, 4\}$".* *Then*

$$\mathbf{FTrafo}(\mathcal{F}) \preccurlyeq \mathbf{FTrafo}(\mathcal{F}_{3,4})$$

Proof. To reduce an instance of **FTrafo**$(\mathcal{F}_n)$ to one of **FTrafo**$(\mathcal{F}_{3,4})$, apply the theorem $\lfloor \frac{n-3}{2} \rfloor$ times.

A form of dimension n has $\frac{n(n+1)}{2} = \Theta(n^2)$ coefficients. Hence an instance of **FTrafo**$(\mathcal{F}_n)$ also has at least that size. Therefore, we have concatenated the reduction of the theorem only polynomially many times, which forms another polynomial-time reduction. □

This corollary is very important for the understanding of the complexity of **Trafo**. It implies that if **Trafo**$(\mathcal{F})$ is hard at all, then **Trafo**$(\mathcal{F}_{3,4})$ necessarily is hard. We now that indefiniteness is necessary for hardness, and the restrictions on the determinant do not seem to make the problem significantly easier. Hence we can argue heuristically in a similar vein for **Trafo** beyond dimension two instead of **Trafo**$(\mathcal{F})$.

Therefore to understand the problem **Trafo**, it suffices to analyse the problem **Trafo**$(\mathcal{F}_{3,4})$.

In the proof of the theorem, we will consider equivalence over local rings $\mathbb{Z}_p$, with p a prime or $\mathbb{Z}_\infty = \mathbb{R}$ (see [5, ch. 7]). f, g are said to belong to the same *genus* if $f \sim_{\mathbb{Z}_p} g$ for all symbols p. Obviously, this is necessary for $f \sim_{\mathbb{Z}} g$ (from now on denoted by $f \sim g$).

Moreover, in [21] a polynomial-time algorithm is presented which given an isotropic integral form f of dimension $n \geq 3$ and given the factorization of its odd squarefree determinant d, computes a matrix $S \in \mathrm{GL}_n\mathbb{Z}$ such that $f S$ has an associated matrix of the shape

$$\begin{pmatrix} 0 & 1 & \\ 1 & 0 & \\ & & A_0 \end{pmatrix}$$

for some matrix A_0 of an $(n-2)$-dimensional form f_0. In other words, the algorithm find a transformation from f to the form

$$2x_1x_2 + f_0(x_3, \ldots, x_n).$$

This works essentially as follows: Using the factorization of d, the algorithm computes a primitive isotropic vector for f, and takes this vector as the first column of S_1. Then by the aid of the Euclidean Algorithm, we enforce that the coefficient a_{12} in the associated matrix of f is the greatest common divisor of $a_{12}, \ldots, a_{1n}$. Then by the analogues of size reduction in lattice theory, we can annullate $a_{13}, \ldots, a_{1n}$. The squarefreeness of the determinant implies $a_{12} = 1$. Similary as for the first line, we obtain $a_{23} = \ldots = a_{2n} = 0$. Finally, a special transformation leads to $a_{22} = 0$.

Proof. (of theorem.) Let (f, g) be an instance of **Trafo**$_n(d)$. By Meyer's theorem [5, sec. 6.1], f is either definite or isotropic since $n \geq 5$. As noted in the introduction, for definite forms the problem can be efficiently solved, so assume indefiniteness. As d is squarefree, f can be efficiently transformed into $f S_1$ with matrix of the shape

$$\begin{pmatrix} 0 & 1 & \\ 1 & 0 & \\ & & A_0 \end{pmatrix}$$

for some matrix A_0 of an $(n-2)$-dimensional form f_0 by [21], using the factorization of $\det f$. Analogously, find a form g_0 for g according to the same procedure. Now by Witt's lemma for p-adic integers [5], $f_0 \sim_{\mathbb{Z}_p} g_0$ for all symbols $p \neq 2$. Since d is odd, it follows that f_0 and g_0 belong to the same genus. But by [5, p. 202f.], we deduce that $f \sim g$ because $\dim f_0 \geq 3$ and d is squarefree. Obviously, S_0 with $f_0 S_0 = g_0$ can be extended to a matrix solving the original problem.   □

5 NP-Hardness

We keep to the base ring $R = \mathbb{Z}$. In [10], it is proved that a decisional variant of the transformation problem is NP-hard under randomized reductions, conditional on what we call the special Cohen-Lenstra heuristic. More precisely, we introduce the following variant of **Trafo**:

> **DITrafo Decisional Interval Transformation Problem**
> *PARAMETERS:* Set $\mathcal{P}$ of properties of quadratic forms.
> *INPUT:* $n \in \mathbb{N}$, n-ary quadratic forms f, g satifying all properties from
> $\mathcal{P}$, matrices $A, B \in (\mathbb{Z} \cup \{\pm\infty\})^{n \times n}$, factorization of $\det f$.
> *DECIDE:* Whether there is $T \in \mathrm{GL}_n\mathbb{Z}$, $A_{ij} \leq T_{ij} \leq B_{ij}$ for all i, j such
>   that $fT = g$.

Note that **DITrafo** is polynomial-time equivalent to the problem of actually computing a transformation with coefficients in the given intervals, by a straightforward divide-and-conquer algorithm.

The *special Cohen-Lenstra Heuristic* (sCLH) builds on a famous conjecture in algebraic number theory, originally published in [6] and [7]. As a tiny (yet important) special case, it implies that class numbers of real quadratic number fields $\mathbb{Q}[\sqrt{p}]$, p prime, equal one with high probability. Very roughly, our assumption stipulates that in this special case, convergence (of frequencies to probability) is not too slow.

Recall that the complexity class RP (*random polynomial time*) consists of all decicion problems for which there is a probabilistic polynomial-time algorithm which accepts every 'yes'-instance with probability $\geq \frac{1}{2}$, and rejects every 'no'-instance.

Theorem 5. *Let $M \in \mathbb{N}$. Let $\mathcal{P}$ consist of the properties $\dim f = 3$, and f indefinite anisotropic for a quadratic form f. If the special Cohen-Lenstra heuristic holds true, then* **DITrafo**$(\mathcal{P})$ *is NP-hard under randomized reductions with one-sided error; more precisely:*

$$NP \subseteq RP^{\mathbf{DITrafo}(\mathcal{P})}. \tag{7}$$

Equation (7) means that every problem from NP can be solved in random polynomial time, given a **DITrafo**($\mathcal{P}$)-oracle. For details on the sCLH, the theorem, and its proof, see [10].

6   Conclusion

We have analyzed the complexity of **Trafo**. Our goal was to find, or exclude, possible key pairs for the identification scheme in Sect. 2.3. We found out that over finite prime fields, the transformation problem is easy, whereas over the rationals, it is only as hard as factoring. This leaves us with the natural choice $R = \mathbb{Z}$ as a base ring. The hardness of **Trafo** over $\mathbb{Z}$ is supported by the NP-hardness results from [10]. Moreover, the concentration of complexity in dimensions 3 and 4 allows for small keys and thus highly efficient cryptographic applications.

References

1. Adleman, L.M., Estes, D.R., McCurley, K.S.: Solving bivariate quadratic congruences in random polynomial time. Mathematics of Computation 48(177), 17–28 (1987)
2. Ajtai, M., Dwork, C.: A public-key cryptosystem with worst- case/average-case equivalence. In: Proceedings of the 29th annual ACM symposium on theory of computing, El Paso, TX, USA, May 4-6, 1997, pp. 284–293. Association for Computing Machinery (1997)
3. Bennett, C.H., Bernstein, E., Brassard, G., Vazirani, U.: Strengths and weaknesses of quantum computing. SIAM Journal of Computing 26(5), 1510–1523 (1997)
4. Blum, L., Blum, M., Shub, M.: A simple unpredictable pseudo- random number generator. SIAM Journal of Computing 15, 364–383 (1986)
5. Cassels, J.W.S.: Rational quadratic forms. Mathematical Society Monographs, vol. 13. Academic Press, London (1978)
6. Cohen, H., Lenstra jun, H.W.: Heuristics on class groups of number elds, Number Theory. In: Proc. Journ. arith., Noordwijkerhout 1983. LNCS, vol. 1068, pp. 33–62. Springer, Heidelberg (1984)
7. Cohen, H., Martinet, J.: Class groups of number elds: Numerical heuristics. Mathematics of Computation 48(177), 123–137 (1987)
8. Crandall, R., Pomerance, C.: Prime numbers: A computational perspective. Springer, Heidelberg (2001)
9. Goldreich, O., Goldwasser, S., Halevi, S.: Public-key cryp- tosystems from lattice reduction problems. In: Kaliski Jr., B.S. (ed.) CRYPTO 1997. LNCS, vol. 1294, pp. 112–131. Springer, Heidelberg (1997)
10. Hartung, R.J., Schnorr, C.-P.: Public key identification based on the equivalence of quadratic forms. In: Kučera, L., Kučera, A. (eds.) MFCS 2007. LNCS, vol. 4708, pp. 333–345. Springer, Heidelberg (2007)
11. Hoffstein, J., Howgrave-Graham, N., Pipher, J., Silverman, J.H., Whyte, W.: NTRUSign: Digital signatures using the NTRU lattice. In: Joye, M. (ed.) CT-RSA 2003. LNCS, vol. 2612, pp. 122–140. Springer, Heidelberg (2003)
12. Hoffstein, J., Pipher, J., Silverman, J.H.: NTRU: A ring-based public key cryptosystem. In: Buhler, J.P. (ed.) ANTS 1998. LNCS, vol. 1423, pp. 267–288. Springer, Heidelberg (1998)

13. Hoffstein, J., Pipher, J., Silverman, J.H.: NSS: an NTRU lattice-based signature scheme. In: Pfitzmann, B. (ed.) EUROCRYPT 2001. LNCS, vol. 2045, pp. 211–228. Springer, Heidelberg (2001)
14. Ivanyos, G., Szántó, Á.: Lattice basis reduction for indefinite forms and an application. Journal on Discrete Mathematics 153(1-3), 177–188 (1996)
15. Kannan, R.: Minkowski's convex body theorem and integer programming. Mathematics of Operations Research 12(3), 415–440 (1987)
16. Khot, S.: Hardness of approximating the shortest vector problem in lattices. Journal of the ACM 52(5), 789–808 (2005)
17. Micciancio, D., Goldwasser, S.: Complexity of lattice problems: a cryptographic perspective. The Kluwer International Series in Engineering and Computer Science, vol. 671. Kluwer Academic Publishers, Dordrecht (March 2002)
18. O'Meara, O.T.: Introduction to quadratic forms, Grundlehren der mathematischen Wissenschaften in Einzeldarstellungen, vol. 117. Springer, Heidelberg (reprinted, 2000)
19. Plesken, W., Pohst, M.E.: Constructing integral lattices with pre- scribed minimum. I, Mathematics of Computation 45, 209–221 (1985)
20. Plesken, W., Souvignier, B.: Computing isometries of lattices. Mathematics of Computation 45, 209–221 (1985)
21. Schnorr, C.-P.: Reduction of quadratic forms reconsidered (preprint, 2004)
22. Simon, D.: Quadratic equations in dimensions 4, 5 and more (preprint, 2005)
23. Simon, D.: Solving quadratic equations using reduced unimodular quadratic forms. Mathematics of Computation 74(251), 1531–1543 (2005)
24. Stehlé, D., Hanrot, G.: Improved analysis of Kannan's shortest lattice vector algorithm. In: Menezes, A.J. (ed.) CRYPTO 2007. LNCS, vol. 4622, pp. 170–186. Springer, Heidelberg (2007)

Towards a Concrete Security Proof of Courtois, Finiasz and Sendrier Signature Scheme

Léonard Dallot

GREYC, UMR 6072, Caen, France
leonard.dallot@info.unicaen.fr

Abstract. Courtois, Finiasz and Sendrier proposed in 2001 a practical code-based signature scheme. We give a rigorous security analysis of a modified version of this scheme in the random oracle model. Our reduction involves two problems of coding theory widely considered as difficult, the Goppa Parametrized Bounded Decoding and the Goppa Code Distinguishing.

1 Introduction

Code-based cryptography was introduced by McElliece [14], two years after the introduction of public key cryptography by Diffie and Hellman [8] in 1976. In 1986 Niederreiter proposed [16] an equivalent code-based cryptosystem [12]. But the first practical code-based signature scheme was proposed in 2001 by Courtois, Finiasz and Sendrier in [7]. It adapts the *Full Domain Hash* approach of Bellare and Rogaway [1] to Niederreiter's encryption scheme. Even if some arguments of its security are given, to the best of our knowledge no formal reductionist security proof was given.

Reductionist security was introduced by Goldwasser and Micali in 1984 [10]. In this approach, a cryptographic scheme is based on one ore more algorithmic problems that are supposed to be hard to solve. The scheme is secure as long as the underlying algorithmic problem is difficult. The works proposed by Bellare and Rogaway in [1] and [2] show the importance of taking into account the tightness of the reduction for practical applications of provable security. The reduction is tight when breaking the scheme leads to solve the considered algorithmic problem with a sufficient probability (ideally one).

Code-based cryptography uses two difficult problems of coding theory. The first one is a NP-complete problem [3], the *Bounded Distance Decoding* problem. The second problem is the *Goppa Code Distinguishing*, widely considered as difficult [18].

We propose a reductionist security proof in the *random oracle model* [1] of a modified version of the signature scheme proposed by Courtois, Finiasz and Sendrier. This proof covers both the security against a key recovering attack usually related to the indistinguishability of a permuted Goppa code, and the security against a decoding attack related to the difficulty of the bounded decoding problem. Using a sequence of games, as proposed by Shoup in [19], we propose an evaluation of the tightness of our reduction.

S. Lucks, A.-R. Sadeghi, and C. Wolf (Eds.): WEWoRC 2007, LNCS 4945, pp. 65–77, 2008.
© Springer-Verlag Berlin Heidelberg 2008

The paper is organized as follows. In Section 2 we present notions on signature schemes. Section 3 presents the security model of our analysis. Some basic concepts of coding theory are given in Section 4. In Section 5, we present a modified version of the signature scheme proposed by Courtois, Finiasz and Sendrier. Section 6 contains our main result (Theorem 1) with its proof.

2   Signature Schemes

A *signature* σ is a bit string dependent of some secret (the secret key SK) only known by the signer and the signed message m. A public value (the public key PK) allows anyone to check the validity of the signature. The following definition is based on [11] and [6].

Definition 1 (Signature Scheme). *A signature scheme S is defined by three algorithms:*

- *$Gen_S(1^\kappa)$, the* key generator algorithm, *is a probabilistic algorithm which, given some security parameter κ outputs a pair of a public key and a secret key (PK, SK) and possibly sets public parameters.*
- *$Sign_S(m, SK)$, the* signing algorithm *takes as input a message m and a secret key SK and ouputs a signature σ.*
- *$Verify_S(m', \sigma', PK)$, the* verification algorithm *takes a message m', a candidate signature σ' and a public key PK as input. It outputs a bit that equals 1 if the signature is accepted and 0 otherwise. We also require that if $\sigma \leftarrow Sign(m, SK)$, then $Verify(m, \sigma, PK) = 1$.*

A very common practice to build signature schemes from a public-key cryptosystem is the "hash and decrypt" paradigm: the message is hashed, possibly some padding is added and this value is decrypted using the private key. This was introduced by Bellare and Rogaway in [1] using RSA [17] as public-key cryptosystem and is known as FDH (*Full Domain Hash*).

3   Security Model

The Random Oracle Model (ROM). A hash function is an application $\mathcal{H} : \{0, 1\}^* \rightarrow \{0, 1\}^n$ where n is fixed. *Preimage resistance* and *second preimage resistance* are generally considered as two prerequisites of a hash function [15].

The random oracle methodology was also introduced by Bellare and Rogaway in [1]. In this model, a hash function is seen as an oracle which produces a uniformly distributed random value for each new query. Provably secure schemes in the random oracle model generally use both algorithmic assumptions and hash functions. Additionally, practical attacks and cryptanalysis of such schemes assume that hash functions are random. The random oracle model guarantees that such an attack cannot be successful unless the algorithmic assumption is false.

Thus, a security proof in the random oracle model gives arguments showing that a real attack — provided that it does not contradict any other assumption of the proof — underlines an undesirable property in the hash function.

Attack Model. The analysis of a cryptographic protocol requires to modelize an adversary, namely setting his goals and his means. Once this is done, the defined adversary serves to solve an algorithmic problem. In other words, the existence of an adversary that is able to break the scheme implies a polynomial algorithm that solves the problem. This approach is comparable to the method of proving that a given decision problem is NP-complete.

The usual attack against a signature scheme is an *existential forgery* in a *chosen message attack* (EF-CMA). A $(\tau, q_{\mathcal{H}}, q_{\Sigma})$-adversary $\mathcal{A}$ knows the public key PK and can obtain $q_{\mathcal{H}}$ hash values for bitstrings of his choice from an (idealised) hash oracle $\mathcal{H}$ and q_{Σ} signatures for messages of its choice from a signature oracle Σ. We impose that each signature query implies a additional hash query. After at most τ processing time, the adversary attempts to output a valid forgery, that is a pair (m^*, σ^*) such that $\mathsf{Verify}(m^*, \sigma^*, PK) = 1$. Obviously, the adversary must output as forgery a signature not obtained form Σ for the message m^*. Another additional hash query can be made to verify the validity of its forgery. Thus the total number of hash queries is $q'_{\mathcal{H}} = q_{\mathcal{H}} + q_{\Sigma} + 1$.

An existential forgery in a chosen message attack can be viewed as a game played between the adversary and a challenger. The challenger runs the generation algorithm, sets the oracles and gives the adversary the public key. He also controls the oracle. Descriptions of games are given as an algorithm the challenger runs. We call the *success of a $(\tau, q_{\mathcal{H}}, q_{\Sigma})$-adversary* $\mathcal{A}$ for an existential forgery in a chosen message attack the probability that $\mathcal{A}$ wins the EF-CMA-game:

$$Succ_S^{EF-CMA}(\mathcal{A}) = \Pr[\mathcal{A} \text{ wins EF-CMA game}].$$

We can now formally define the security of a signature scheme.

Definition 2. *A signature scheme S is $(\epsilon, \tau, q_{\mathcal{H}}, q_{\Sigma})$-EF-CMA-secure if for any $(\tau, q_{\mathcal{H}}, q_{\Sigma})$-adversary $\mathcal{A}$:*

$$Succ_S^{EF-CMA}(\mathcal{A}) \leq \epsilon.$$

Remark 1. EF-CMA security includes security against a *key recovering attack*. Indeed, an adversary who can recover the private key of the signer can easily compute a signature for a message of its choice and then outputs this message and its signature as an existential forgery.

4   Coding Theory Background

Let $\mathbb{F}_q$ be the field with q elements. A (n, k)-*code* $\mathcal{C}$ is a linear subspace of dimension k of the linear space $\mathbb{F}_q^n$. Elements of $\mathbb{F}_q^n$ are called *words* and elements of $\mathcal{C}$ are *codewords*. A code is usually given in the form of a $(n-k) \times n$ *parity check matrix* H. The codewords of $\mathcal{C}$ are words x that satisfy $Hx^T = 0$. A *syndrome* $s \in \mathbb{F}_q^{n-k}$ is a vector $s = Hx^T$ for a word x. The *Hamming weight* of a word x denoted by $\mathsf{wt}(x)$ is the number of non-zero positions. A syndrome s is said to be *decodable* according to a t-error correcting code if there exists a word $x \in \mathbb{F}_q^n$

such that $Hx^T = s$ and $\text{wt}(x) \leq t$. We recall that decoding a syndrome s is retrieving such a word x.

Goppa Codes are subfield subcodes of particular alternant codes [13]. They are widely used in code-based cryptography. For given integers m and t Goppa codes are of length $n = 2^m$, of dimension $k = n - mt$ and are t-correcting. The density of decodable syndromes is approximately $\frac{1}{t!}$ [7]. We denote by $Decode_H$ the decoding algorithm associated with a Goppa Code of parity check matrix H.

In order to establish the security of code-based cryptographic schemes, we shall consider the *Goppa Parameterized Bounded Decoding problem* (GPBD) introduced in [9] and [18]. This problem is a variant of a NP-hard problem of coding theory, namely the *Bounded Decoding problem* [3].

Definition 3 (Goppa Parameterized Bounded Decoding problem (GPBD) [9]).

Input: A $(n - k) \times n$ binary matrix H and a syndrome $s \in \mathbb{F}_2{}^{n-k}$
Ouput: A word $e \in \mathbb{F}_2^n$ such that $\text{wt}(e) \leq \frac{n-k}{\log_2 n}$ and $He^T = s$

The resolution of this problem can be viewed as a game played between a challenger and decoder $\mathcal{Dec}$. The challenger gets a random binary (n, k)-code $\mathcal{C}$ and a random syndrome $s \in \mathbb{F}_2{}^{n-k}$. It gives the parity check matrix H of $\mathcal{C}$ and the syndrome s to the decoder. $\mathcal{Dec}$ outputs a word e. If the syndrome of e is equal to s and e satisfy the weight property, then $\mathcal{Dec}$ wins the game. A description of GPBD game is given in Fig. 2. The notations $x \xleftarrow{R} D$ denotes a random selection of x over a given distribution D. If D is a set, $\xleftarrow{R}$ denotes an uniformly distributed selection over the set. We call the *success of the decoder $\mathcal{Dec}$* the probability that it wins the game:

$$Succ^{GPBD}(\mathcal{Dec}) = \Pr[\mathcal{Dec} \text{ wins GPBD game}]$$

Definition 4. *The Goppa Parameterized Bounded Decoding problem is said to be (τ, ϵ)-hard if for any decoder $\mathcal{D}$ running in time at most τ we have $Succ^{GPBD}(\mathcal{D}) \leq \epsilon$.*

Input: An adversary $\mathcal{A}$
1 $(SK, PK) \leftarrow \text{Gen}_S(1^\kappa)$;
2 Set the oracles $\mathcal{H}$ and Σ;
3 $(m^*, \sigma^*) \leftarrow \mathcal{A}^{\Sigma, \mathcal{H}}(PK)$;
4 **if** $\text{Verify}_S(m^*, \sigma^*, PK) = 1$ **and** Σ *did not provide* σ^* **then**
5 $\quad |\quad \mathcal{A}$ wins the game
6 **else**
7 $\quad |\quad \mathcal{A}$ loses the game
8 **end**

Fig. 1. EF-CMA Game

```
    Input: A decoder Dec
1  (C, H) ←ᴿ Binary(n, k);
2  s ←ᴿ F₂ⁿ⁻ᵏ;
3  e ← Dec(H);
4  if  Heᵀ = s and wt(e) ≤ (n−k)/log₂ n  then
5  |    Dec wins the game
6  else
7  |    Dec loses the game
8  end
```

Fig. 2. GPBD Game

Since GPBD problem is stated for random codes and since practical code-based cryptography generally uses Goppa codes, we also have to consider the *Goppa Code Distinguishing* problem (GD) presented in [18]. A distinguisher $\mathcal{D}$ for a permuted Goppa Code is an algorithm which takes as input a parity check matrix H and outputs a bit. $\mathcal{D}$ outputs 1 with probability $\Pr[H \xleftarrow{R} \mathsf{Goppa}(n,k) : \mathcal{D}(H) = 1]$ if H is a random binary parity check matrix of a Goppa code $\mathsf{Goppa}(n,k)$ and outputs 1 with probability $\Pr[H \xleftarrow{R} \mathsf{Binary}(n,k) : \mathcal{D}(H) = 1]$ if H is a random binary matrix $\mathsf{Binary}(n,k)$. We call the *advantage of a distinguisher* $\mathcal{D}$ the following quantity:

$$Adv^{GD}(\mathcal{D}) =$$
$$\left| \Pr[H \xleftarrow{R} \mathsf{Goppa}(n,k) : \mathcal{D}(H) = 1] - \Pr[H \xleftarrow{R} \mathsf{Binary}(n,k) : \mathcal{D}(H) = 1] \right|$$

Definition 5. *The Goppa Code Distinguishing is said to be (τ, ϵ)-hard if for any distinguisher $\mathcal{D}$ running in time at most τ we have $Adv^{GD}(\mathcal{D}) \leq \epsilon$.*

5   Code-Based Signatures: Courtois, Finiasz and Sendrier Scheme

Courtois, Finiasz and Sendrier proposed in [7] the first practical signature scheme based on coding theory. The FDH approach assumes that all the hash values can be inverted by decryption. But in code-based cryptography, only decodable syndromes can be decrypted. To overcome this difficulty, the authors proposed to adapt the FDH approach (see Section 2) to permit multiple hash values for the same message by concatenating to the message a counter before hashing. If the decryption (*i.e.* the decoding) fails, the counter is incremented until a decodable hash value is found.

We propose to replace the counter by a random value uniformly distributed over $\{1, \ldots, 2^{n-k}\}$. The Gen and Verify algorithms remains the same as in the original scheme CFS (called CFS₁ in the original paper).

- $\text{Gen}_{\text{mCFS}}(1^{\kappa})$: Select n, k and t according to κ. Pick a random parity check matrix H_0 of a (n, k)-binary Goppa code $\mathcal{C}_0$ decoding t errors. This code remains secret. The public code is obtained by randomly permuting the coordinates of $\mathcal{C}_0$ and then choosing a random parity check matrix. Choose a random $(n-k) \times (n-k)$ non-singular matrix U, a random $n \times n$ permutation matrix P and a hash function $h : \{0,1\}^* \longrightarrow \mathbb{F}_2^{n-k}$. The public key is $H = U H_0 P$ and the private key is (U, H_0, P). Set $t = \frac{n-k}{\log_2 n}$.

- $\text{Sign}_{\text{mCFS}}(m, H_0)$:
  1. $i \xleftarrow{R} \{1, \ldots, 2^{n-k}\}$
  2. $x' = \mathcal{D}ecode_{H_0}\left(U^{-1} h(m\|i)\right)$
  3. If no x' was found go to 2
  4. output $(i, x'P)$

- $\text{Verify}_{\text{mCFS}}(m, x', i', H)$: compute $s' = H x'^T$ and $s = h(m\|i)$. The signature is valid if s and s' are equals.

Remark 2. The modified scheme uses $n - k$ bits (144 bits with the original parameters) to store the counter instead of $\log_2 t!$ bits in average (around 19 bits) in the original construction. The reason of our modification is that the counter gives to an adversary a piece of information he may exploit in an attack. If i is the signature counter, no hash values for counter $j < i$ are decodable.

6 Proving Security of mCFS

To prove the security of the mCFS scheme, we will use the methodology of Shoup [19] by producing a sequence of games relating the EF-CMA game (Fig. 1) to the GPBD game (Fig. 2). Each game is a slight modification of the preceding game in a way that the difference between two games can be evaluate. Thus the quality of the reduction can be easily quantified. Since the challenger can control the oracles he is able to produce simulations of oracles that force the adversary to solve the GPBD game. We state the following:

Theorem 1. *Suppose that the* Goppa Parametrized Bounded Decoding Problem *and the* Goppa Code Distinguishing *are respectively* $(\tau_{GPBD}, \epsilon_{GPBD})$ *and* $(\tau_{GD}, \epsilon_{GD})$-*hard. Then the modified* CFS *scheme is* $(\epsilon, \tau, q_{\mathcal{H}}, q_{\Sigma})$-EF-CMA-*secure in the* random oracle model, *where:*

$$\epsilon = (q_{\mathcal{H}} + q_{\Sigma} + 1)\epsilon_{GPBD} + \epsilon_{GD} + 2 - (1 - \frac{1}{2^{n-k}})^{q_{\mathcal{H}} + q_{\Sigma} + 1} - (1 - \frac{q_{\Sigma}}{2^{n-k}})^{q_{\mathcal{H}}}$$

and

$$\tau \geq \tau_{GPBD} - (q_{\mathcal{H}} + q_{\Sigma} + 1) \cdot T_s(n, k)$$

where $T_s(n, k)$ *is the syndrome computation time of a* (n, k)-*Goppa Code.*

Proof. Let $\mathcal{A}$ be a $(\tau, q_{\mathcal{H}}, q_{\Sigma})$-adversary against the modified mCFS scheme. Let Game 0 (Fig. 3) be the standard EF-CMA game adapted to the scheme. We

> **Input**: An adversary $\mathcal{A}$
> 1 $(H_0, U, P, H) \leftarrow \mathsf{Gen}_{\mathsf{mCFS}}(n, k)$;
> 2 Set the oracles $\mathcal{H}$ and Σ;
> 3 $(m^*, \sigma^*, i^*) \leftarrow \mathcal{A}^{\Sigma, \mathcal{H}}(H)$;
> 4 **if** $\begin{cases} \mathcal{H}(m^* i^*) = H\sigma^{*T} \\ \mathsf{wt}(\sigma^*) \leq t \end{cases}$ **and** Σ did not provide σ^* **then**
> 5 $\quad |\quad \mathcal{A}$ wins the game
> 6 **else**
> 7 $\quad |\quad \mathcal{A}$ loses the game
> 8 **end**

Fig. 3. game 0: mCFS EF-CMA game

denote by $\Pr[S_i]$ the probability that $\mathcal{A}$ wins the game i. We have $Pr[S_0] = Succ_{\mathsf{mCFS}}^{EF-CMA}(\mathcal{A})$.

To simplify the proof, we will consider that hash queries are made on pairs (m, i) of messages and indexes. Oracles $\mathcal{H}$ and Σ maintain lists $\Lambda_{\mathcal{H}}$ and Λ_{Σ} respectively of queries with the corresponding output values. Since the oracles are controlled by the challenger, each simulated oracle may access these lists. The list $\Lambda_{\mathcal{H}}$ is extended to store an additional decoding value. For any message m and any counter i, $\Lambda_{\mathcal{H}}(m, i) = (s, x)$ where s is the output syndrome and x its decoding. $\Lambda_{\Sigma}(m)$ is then equals to (i, x). We also use an additional list Λ which applied to m return a counter $\Lambda(m)$. If there is no value associated with an entry in a list, we denote the output by $\perp$.

Game 1. In this game, the challenger replaces the hash oracle $\mathcal{H}$ by a simulation $\mathcal{H}'$ (Fig. 4). $\mathcal{H}'$ uses the list Λ to fix for each message which counters leads to a decodable syndrome (Fig. 4, Lines 1 to 3). Thus, they are two situations for any query (m, j): either $j \neq \Lambda(m)$ or $j = \Lambda(m)$.

When $j \neq \Lambda(m)$ $\mathcal{H}'$ has the same behaviour as a random oracle (Fig. 4, Lines 5 to 11). When $j = \Lambda(m)$ $\mathcal{H}'$ builds a t-decodable syndrome and stores its decoding value in the list $\Lambda_{\mathcal{H}}$: it first gets a random word x of weight t in $\mathbb{F}_2^n$ (Fig. 4, Line 13), then it computes its syndrome s as output (Fig. 4, Line 14). $\mathcal{H}'$ stores the values s and x into $\Lambda_{\mathcal{H}}$. Of course, $\mathcal{H}'$ checks if he has an ouput value stored in his list for the query (Fig. 4, Lines 6 and 12).

At the end of the simulation, the oracle $\mathcal{H}'$ has produced $q_{\mathcal{H}} + q_{\Sigma} + 1$ syndromes. Some of them are produced by the modified part of the oracle (when $j = \Lambda(m)$). Let M be the random variable that represents the number of syndromes produced by the modified part.

$$\begin{aligned} \Pr[S_1] &= \Pr[(S_1 \cap (M = 0)) \cup (S_1 \cap (M > 0))] \\ &\leq \Pr[S_1 \cap (M = 0)] + \Pr[S_1 \cap (M > 0)] \\ &\leq \Pr[S_1 \cap (M = 0)] + \Pr[M > 0] \end{aligned}$$

$\Pr[S_1 \cap (M = 0)]$ corresponds to the case where the adversary wins Game 1 with syndrome produced by a random oracle. This is exactly Game 0. Hence $\Pr[S_1 \cap (M = 0)] = \Pr[S_0]$.

Input: A pair (m, j)
Output: A syndrome s
1 if $\Lambda(m) = \perp$ then
2 $\quad | \quad \Lambda(m) \xleftarrow{R} \{1, \ldots, 2^{n-k}\};$
3 end
4 $(s, x) \leftarrow \Lambda_{\mathcal{H}}(m, j);$
5 if $j \neq \Lambda(m)$ then
6 $\quad |$ if $s = \perp$ then
7 $\quad | \quad | \quad s \xleftarrow{R} \mathbb{F}_2^{n-k};$
8 $\quad | \quad | \quad \Lambda_{\mathcal{H}}(m, j) \leftarrow (s, \perp);$
9 $\quad |$ end
10 $\quad |$ return $\mathcal{H}(m, j) = s;$
11 else
12 $\quad |$ if $s = \perp$ then
13 $\quad | \quad | \quad x \xleftarrow{R} \{w \in \mathbb{F}_2^n | \mathsf{wt}(w) \leq t\};$
14 $\quad | \quad | \quad s \leftarrow Hx^T;$
15 $\quad | \quad | \quad \Lambda_{\mathcal{H}}(m, j) \leftarrow (s, x);$
16 $\quad |$ end
17 $\quad |$ return $\mathcal{H}(m, j) = s;$
18 end

Fig. 4. $\mathcal{H}'$: simulation of $\mathcal{H}$ (Game 1)

Input: A message m
Output: A signature (i, σ)
1 if $\Lambda(m) = \perp$ then
2 $\quad | \quad \Lambda(m) \xleftarrow{R} \{1, \ldots, 2^{n-k}\};$
3 end
4 $\mathcal{H}'(m, \Lambda(m));$
5 $(s, x) \leftarrow \Lambda_{\mathcal{H}}(m, \Lambda(m));$
6 $\Lambda(m) \leftarrow \perp;$
7 return $\Sigma(m) = (i, x);$

Fig. 5. Σ': simulation of Σ (Game 1)

M respects a binomial distribution of parameters $\frac{1}{2^{n-k}}$ and $q_{\mathcal{H}} + q_{\Sigma} + 1$ and then $\Pr[M > 0] = 1 - (1 - \frac{1}{2^{n-k}})^{q_{\mathcal{H}} + q_{\Sigma} + 1}$. It follows:

$$|\Pr[S_1] - \Pr[S_0]| \leq 1 - (1 - \frac{1}{2^{n-k}})^{q_{\mathcal{H}} + q_{\Sigma} + 1}$$

Game 2. In this game, the challenger replaces the signature oracle by a simulation Σ' (Fig. 5). Since Σ' queries $\mathcal{H}'$ on $(m, \Lambda(m))$, $\mathcal{H}'$ stores the decoding value of its output. Thus Σ' no more need the private key to produce signatures. Σ' also deletes $\Lambda(m)$ in order that two different signature queries for the same message does not produce the same signature, according to the modified scheme.

```
    Input: A parity check matrix H
    Output: A bit b
1   t ← n-k / log₂ n ;
2   Set the oracles H' and Σ';
3   (m*, σ*, i*) ← A^(Σ',H')(H);
    if   { H'(m*‖i*) = Hσ*^T   and Σ' did not provide σ* then
4        { wt(σ*) ≤ t
5   |    output 1
6   else
7   |    output 0
8   end
```

Fig. 6. $\mathcal{D}(H)$ (game 4)

$\mathcal{A}$ may query $\mathcal{H}'$ on messages he already queried to Σ'. Then, $\mathcal{H}'$ return a syndrome produced by the modified part of the oracle. This happens with probability at most $\frac{q_\Sigma}{2^{n-k}}$. Let S be the number of such syndromes.

$$\Pr[S_2] = \Pr\left[(S_2 \cap (S = 0)) \cup (S_2 \cap (S > 0))\right]$$
$$\leq \Pr[S_2 \cap (S = 0)] + \Pr[S_2 \cap (S > 0)]$$
$$\leq \Pr[S_1] + \Pr[S > 0]$$
$$\leq \Pr[S_1] + 1 - \left(1 - \frac{q_\Sigma}{2^{n-k}}\right)^{q_\mathcal{H}}$$

Game 3. In this game the challenger replaces the generation algorithm Gen_{mCFS} by a random selection of a parity check matrix of a binary Goppa code. This code is used as the public key. Since neither the hash oracle or the signature oracle no more use the private key and the hash function, the simulation is not altered and then:

$$\Pr[S_3] = \Pr[S_2].$$

Game 4. In this game, the challenger replaces the random binary Goppa code by a random binary code. Then we can build the distinguisher presented Fig. 6. If H is a permuted binary Goppa code, $\mathcal{D}$ proceeds as Game 3 and therefore

$$\Pr[H \xleftarrow{R} \mathsf{Goppa}(n,k) : \mathcal{D}(H) = 1] = \Pr[S_3].$$

If H is a random binary code, $\mathcal{D}$ proceeds as Game 4 and therefore

$$\Pr[H \xleftarrow{R} \mathsf{Binary}(n,k) : \mathcal{D}(H) = 1] = \Pr[S_4].$$

Then, $Adv^{GD}(\mathcal{D}) = |\Pr[S_3] - \Pr[S_4]|$. Since we suppose the distinguish of permuted Goppa code problem as $(\tau_{GD}, \epsilon_{GD})$-hard

$$|\Pr[S_3] - \Pr[S_4]| \leq \epsilon_{GD}$$

Game 5. In this game (Fig. 7), the challenger modify the winning condition. This game is conditioned by the adversary making its forgery on a particular

---

Input: An adversary $\mathcal{A}$

1　$c \xleftarrow{R} \{1, \ldots, q_{\mathcal{H}} + q_{\Sigma} + 1\}$;

2　$H^* \xleftarrow{R} \text{Binary}(n, k)$;

3　$t \leftarrow \frac{n-k}{\log_2 n}$;

4　Set the oracles $\mathcal{H}'$ and Σ';

5　$(m^*, \sigma^*, i^*) \leftarrow \mathcal{A}^{\Sigma', \mathcal{H}'}(H^*)$;

6　**if** $\begin{cases} \mathcal{H}'(m^*, i^*) = H^* \sigma^{*T} \\ wt(\sigma^*) \leq t \end{cases}$ *and* $\begin{cases} \Sigma' \text{ did not provide } \sigma^* \\ c\text{-th query to } \mathcal{H}' \text{ was } (m^*, i^*) \end{cases}$ **then**

7　| $\mathcal{A}$ wins the game

8　**else**

9　| $\mathcal{A}$ loses the game

10　**end**

Fig. 7. Game 5

Input: An adversary $\mathcal{A}$

1　$c \xleftarrow{R} \{1, \ldots, q_{\mathcal{H}} + q_{\Sigma} + 1\}$;

2　$H^* \xleftarrow{R} \text{Binary}(n, k)$;

3　$t \leftarrow \frac{n-k}{\log_2 n}$;

4　$s^* \xleftarrow{R} s \in \mathbb{F}_2^{n-k}$;

5　Set the oracles $\mathcal{H}'$ and Σ';

6　$(m^*, \sigma^*, i^*) \leftarrow \mathcal{A}^{\Sigma', \mathcal{H}'}(H^*)$;

7　**if** $\begin{cases} \mathcal{H}'(m^*, i^*) = H^* \sigma^{*T} \\ wt(\sigma^*) \leq t \end{cases}$ *and* $\begin{cases} \Sigma' \text{ did not provide } \sigma^* \\ c\text{-th query to } \mathcal{H}' \text{ was } (m^*, i^*) \end{cases}$ **then**

8　| $\mathcal{A}$ wins the game

9　**else**

10　| $\mathcal{A}$ loses the game

11　**end**

Fig. 8. Game 6

---

hash query: the challenger first gets a random $c \xleftarrow{R} \{1, \ldots, q_{\mathcal{H}} + q_{\Sigma} + 1\}$. $\mathcal{A}$ wins the game if, in addition to the preceding winning conditions, the c-th query to $\mathcal{H}'$ was made on (m^*, i^*).

This event, independent from the choice of the adversary, has probability $\frac{1}{q_{\mathcal{H}} + q_{\Sigma} + 1}$ and then, we have:

$$\Pr[S_5] = \frac{\Pr[S_4]}{q_{\mathcal{H}} + q_{\Sigma} + 1}$$

Game 6. In this game (Fig. 8), the challenger modify the hash oracle to output a random syndrome s^* to the c-th query. The probability space is not modified and then $\Pr[S_6] = \Pr[S_5]$.

With the restriction made in game 5, the challenger knows $\mathcal{A}$ will make its forgery on the result s^* of the c-th query (m^*, i^*) to the hash oracle. Hence if $\mathcal{A}$ wins the game we have:

$$\begin{cases} H\sigma^{*T} = \mathcal{H}'(m^*, i^*) = s^* \\ \mathsf{wt}(\sigma^*) \leq t = \frac{n-k}{\log_2 n} \end{cases}$$

Then the adversary wins the GPBD game and $Succ^{GPBD}(\mathcal{A}) = \Pr[S_6]$. From the hypothesis, $Succ^{GPBD}(\mathcal{A}) \leq \epsilon_{GPBD}$ and finally,

$$\Pr[S_6] \leq \epsilon_{GPBD}$$

Sum up. From this sequence of games we have:

1. $\Pr[S_0] = Succ_{mCFS}^{EF-CMA}(\mathcal{A})$
2. $|\Pr[S_0] - \Pr[S_1]| \leq 1 - \left(1 - \frac{1}{2^{n-k}}\right)^{q_{\mathcal{H}}+q_{\Sigma}+1}$
3. $|\Pr[S_1] - \Pr[S_2]| \leq 1 - \left(1 - \frac{q_{\Sigma}}{2^{n-k}}\right)^{q_{\mathcal{H}}}$
4. $\Pr[S_2] = \Pr[S_3]$
5. $|\Pr[S_4] - \Pr[S_3]| \leq \epsilon_{GD}$
6. $\frac{\Pr[S_4]}{q_{\mathcal{H}}+q_{\Sigma}+1} = \Pr[S_5] = \Pr[S_6] = Succ^{GPBD}(\mathcal{A}) \leq \epsilon_{GPBD}$

Using triangular inequality on 2 and 3, we obtain

$$|\Pr[S_0] - \Pr[S_2]| \leq |\Pr[S_0] - \Pr[S_1]| + |\Pr[S_1] - \Pr[S_2]|$$
$$\leq 2 - \left(1 - \frac{1}{2^{n-k}}\right)^{q_{\mathcal{H}}+q_{\Sigma}+1} - \left(1 - \frac{q_{\Sigma}}{2^{n-k}}\right)^{q_{\mathcal{H}}}$$

Let $f(n, k, q_{\mathcal{H}}, q_{\Sigma}) = 2 - \left(1 - \frac{1}{2^{n-k}}\right)^{q_{\mathcal{H}}+q_{\Sigma}+1} - \left(1 - \frac{q_{\Sigma}}{2^{n-k}}\right)^{q_{\mathcal{H}}}$. From 4 and 5, we have:

$$|\Pr[S_0] - \Pr[S_4]| \leq |\Pr[S_0] - \Pr[S_2]| + |\Pr[S_2] - \Pr[S_4]|$$
$$\leq \epsilon_{GD} + f(n, k, q_{\mathcal{H}}, q_{\Sigma}).$$

Since $\Pr[S_4] = (q_{\mathcal{H}} + q_{\Sigma} + 1)\Pr[S_5] = (q_{\mathcal{H}} + q_{\Sigma} + 1)\Pr[S_6]$,

$$|\Pr[S_0] - (q_{\mathcal{H}} + q_{\Sigma} + 1)\Pr[S_6]| \leq \epsilon_{GD} + f(n, k, q_{\mathcal{H}}, q_{\Sigma})$$

Finally, since $\Pr[S_6] \leq \epsilon_{GPBD}$ and $\Pr[S_0] = Succ_{mCFS}^{EF-CMA}(\mathcal{A})$, we have:

$$Succ_{mCFS}^{EF-CMA}(\mathcal{A}) \leq (q_{\mathcal{H}} + q_{\Sigma} + 1)\epsilon_{GPBD} + \epsilon_{GD} + f(n, k, q_{\mathcal{H}}, q_{\Sigma})$$

The running time of the simulation is the running time of $\mathcal{A}$ and the time needed to compute the $q_{\mathcal{H}}+q_{\Sigma}+1$ hash values. This time is at most $q_{\mathcal{H}}+q_{\Sigma}+1$ syndrome computations. This gives the formula for τ.

6.1   Discussion

The basic idea of our proof resembles to the proof of Bellare and Rogaway [1] that FDH RSA ist secure in the ROM: define the simulated hash value as the application of the one-way function of a randomly chosen signature.

The above reduction is clearly not tight: the probability of solving the GPBD problem severely decreases when the number of queries $q_{\mathcal{H}}$ and q_{Σ} increases. The most restrictive parameter is $q_{\mathcal{H}}$ since the number q_{Σ} of signatures can be arbitrary bounded while the number $q_{\mathcal{H}}$ of hash queries is unbounded. This cost in the reduction is essentially due to the bet made on the query used for the forgery.

To tighten the proof of FDH RSA, Coron [6] proposed a simulation of the hash oracle which "hides" with a given probability the RSA challenge into the answers of the oracle. This hiding is realised by multiplicating the application of the one-way function and the challenge. This leads to a new random instance of RSA. Then, the final probability that breaking RSA FDH scheme leads to solve RSA only depends of the number q_{Σ} of signature queries. Unfortunately this approach cannot be used in our proof since building a new challenge for GPBD from another syndrome may increase the weight of the corresponding signature.

7   Conclusion

We have studied the security of a modified version of the first practical signature scheme based on coding theory proposed by Courtois, Finiasz and Sendrier in 2001. This security proof relies on the difficulty of the *Goppa Parametrized Bounded Decoding* (GPBD) problem and the *Goppa Code Distinguishing* (GD) problem. Thanks to the security proof of the *mCFS* scheme, it can be possible to build provable special signature schemes such as undeniable [5] or blind [4] signature schemes.

Unfortunately the reduction is not tight and finding a better reduction or proving the optimality of our proof (*i.e.* prove there can not exist a better one) remains an open problem.

Acknowledgements. I would like to thank D. Vergnaud for the idea of this paper, F. Laguillaumie and A. Otmani for usefull discussions.

References

1. Bellare, M., Rogaway, P.: Random oracles are practical: A paradigm for designing efficient protocols. In: ACM Conference on Computer and Communications Security, pp. 62–73 (1993)
2. Bellare, M., Rogaway, P.: The exact security of digital signatures – how to sign with rsa and rabin. In: Maurer, U.M. (ed.) EUROCRYPT 1996. LNCS, vol. 1070. pp. 399–416. Springer, Heidelberg (1996)
3. Berlekamp, E.R., McEliece, R.J., van Tilborg, H.C.: On the inherent intractability of certain coding problems. IEEE Trans. Inform. Th. 24 (1978)
4. Chaum, D.: Blind signatures for untraceable payments. In: Advances in Cryptology – CRYPTO 1982, Lecture Notes Computer Science, p. 153. Springer, Heidelberg (1982)
5. Chaum, D., van Anderpen, H.: Undeniable signatures. In: Brassard, G. (ed.) CRYPTO 1989. LNCS, vol. 435, pp. 212–216. Springer, Heidelberg (1990)

6. Coron, J.S.: On the exact security of full domain hash. In: Bellare, M. (ed.) CRYPTO 2000. LNCS, vol. 1880, pp. 229–236. Springer, Heidelberg (2000)
7. Courtois, N., Finiasz, M., Sendrier, N.: How to achieve a McEliece-based digital signature scheme. In: Boyd, C. (ed.) ASIACRYPT 2001. LNCS, vol. 2248, pp. 157–174. Springer, Heidelberg (2001)
8. Diffie, W., Hellman, M.: New directions in cryptography. IEEE Trans. Inform. Th. 22(6), 644–654 (1976)
9. Finiasz, M.: Nouvelles constructions utilisant des codes correcteurs d'erreurs en cryptographie à clef publique. PhD thesis, INRIA – Ecole Polytechnique (October 2004) (in French)
10. Goldwasser, S., Micali, S.: Probabilistic encryption. Journal of Computer and Systems Sciences 28(2), 270–299 (1984)
11. Goldwasser, S., Micali, S., Rivest, R.L.: A digital signature scheme secure against adaptive chosen-message attacks. SIAM Journal on Computing 17(2), 281–308 (1988)
12. Li, Y.X., Deng, R.H., Wang, X.M.: On the equivalence of McEliece's and Niederreiter's public-key cryptosystems. IEEE Trans. Inform. Th. 40(1), 271–273 (1994)
13. MacWilliams, F.J., Sloane, N.J.A.: The Theory of Error-Correcting Codes. North-Holland mathematical library, Amsterdam (1977)
14. McEliece, R.J.: A public-key cryptosystem based on algebraic coding theory. Technical report, DSN Progress report # 42-44, Jet Propulsion Laboratory, Pasadena, Californila (1978)
15. Menezes, A.J., Vanstone, S.A., van Oorschot, P.C.: Handbook of Applied Cryptography. CRC Press, Inc., Boca Raton (1996)
16. Niederreiter, H.: Knapsack-type cryptosystems and algebraic coding theory. Problems of Control and Information Theory 15(2), 159–166 (1986)
17. Rivest, R., Shamir, A., Adleman, L.: A method for obtaining digital signatures and public key cryptosystems. CACM 21 (1978)
18. Sendrier, N.: Cryptosystèmes à clé publique basés sur les codes correcteurs d'erreurs. Habilitation à diriger les recherches, Université Pierre et Marie Curie, Paris 6, Paris, France (March 2002) (in French)
19. Shoup, V.: Sequences of games: a tool for taming complexity in security proofs (manuscript, November 2004) (revised, May 2005; January 2006)

Cryptanalysis of MOR and Discrete Logarithms in Inner Automorphism Groups

Anja Korsten

Universität Tübingen, Wilhelm-Schickard-Institut für Informatik,
Sand 14, 72076 Tübingen, Germany
akorsten@informatik.uni-tuebingen.de

Abstract. The MOR cryptosystem was introduced in 2001 as a new public key cryptosystem based on non-abelian groups. This paper demonstrates that the complexity of breaking MOR based on groups of the form $GL(n,q) \times_\theta \mathcal{H}$ ($\mathcal{H}$ a finite abelian group) is (with respect to polynomial reduction) not higher than the complexity of the discrete logarithm problem in small extension fields of $\mathbb{F}_q$. Additionally we consider the construction of a generic attack on MOR.

Keywords: public key cryptography, non-abelian group, MOR cryptosystem, discrete logarithm problem, inner automorphism problem.

1 Introduction

In recent years there has been considerable interest in public key cryptosystems on non-abelian groups, e.g. braid groups or linear groups. Paeng et al. introduced the MOR cryptosystem in 2001, see [7]. This ElGamal-type public key cryptosystem uses the fact that there is no subexponential-time algorithm known to solve the discrete logarithm problem in the inner automorphism group $Inn(G)$ of a non-abelian group G.

Paeng et al. propose the semidirect product $SL(2,p) \times_\theta \mathbb{Z}_p$ as a group for the MOR cryptosystem, but MOR on this group is insecure, as shown by Tobias in [11,12,13]. The presented attacks enable an adversary to derive significant parts of the plaintext or even the secret encryption exponent. These attacks use special properties of $SL(2,p)$ and do not work for semidirect products of $GL(n,q)$ by an arbitrary abelian group. However, we present a ciphertext-only attack which shows that the security of MOR on such groups solely depends on the difficulty of the discrete logarithm problem in small extension fields of $\mathbb{F}_q$. Furthermore we use the idea of the attack to develop a generic reduction of the security of MOR to the underlying group G.

2 The MOR Cryptosystem

Let G be a non-abelian finite group with a set of generators $\{\gamma_1, \ldots, \gamma_l\}$ for $l \in \mathbb{N}$. We assume that the *representation problem* in G is efficiently solvable,

S. Lucks, A.-R. Sadeghi, and C. Wolf (Eds.): WEWoRC 2007, LNCS 4945, pp. 78–89, 2008.

i.e. there exists a polynomial time algorithm that computes the representation of an element of G as a product of the generating elements.

We will consider the group $Inn(G)$ of inner automorphisms of G. The elements of this group are of the form

$$I_g : G \to G, \ x \mapsto gxg^{-1}$$

for $g \in G$. Note that $(I_g)^a = I_{g^a}$ for $a \in \mathbb{N}$, and that $I_g = id_G$ if and only if $g \in Z(G)$. It follows that the inducing element of an inner automorphism is not necessarily unique, but we have $I_g = I_h$ iff $g \in h \cdot Z(G)$.

In the setting of the MOR cryptosystem the inner automorphism induced by an element $g \in G$ will be represented as the set

$$I_g = \{I_g(\gamma_i) : 1 \leq i \leq l\},$$

where each of the $I_g(\gamma_i)$ is again represented as a product of the γ_i's. With this representation it is possible to hide the element g in I_g. This is essential for MOR which is described as follows, see also [7,11].

Public Key Encryption Scheme: **MOR**

Key generation: Bob chooses an arbitrary element $g \in G \backslash Z(G)$: the private key is $a \bmod ord(I_g)$, the public key is (I_g, I_{g^a}).

Encryption:

1. Alice takes a plaintext $m \in G \backslash Z(G)$.
2. Alice chooses an arbitrary private encryption exponent $b \bmod ord(I_g)$ and computes $(I_{g^a})^b = I_{g^{ab}}$.
3. Alice computes $E := I_{g^{ab}}(m)$ and $\varphi := (I_g)^b$.
4. Alice sends (E, φ) to Bob.

Decryption: Bob receives a ciphertext (E, φ) and uses his private key a to compute

$$\varphi^{-a}(E) = I_{g^{-ab}}(E) = I_{g^{-ab}}(I_{g^{ab}}(m)) = m.$$

From this description it is immediately clear that MOR is broken if we are able to compute a from the public key (I_g, I_{g^a}), i.e. by solving a discrete logarithm problem in $< I_g >$.

Definition 1. *The discrete logarithm problem* (DLP) *is defined as follows: Given* $g \in G$ *and* $h \in < g >$ *find* $a \in \mathbb{N}$ *such that* $g^a = h$. *We write* $a \in$ DLP(g, h).

Due to the representation of inner automorphism used for MOR, we consider two other computational problems:

The general discrete logarithm problem (gDLP) *is defined as follows: Given* $g \in G$ *and* $h \in < g > \cdot Z(G)$ *find* $a \in \mathbb{N}$ *and* $z \in Z(G)$ *such that* $g^a = zh$. *We write* $a \in$ gDLP$_G(g, h)$.

The inner automorphism problem (IAP) *is defined as follows: Given* $I_g \in Inn(G)$ *by* $I_g = \{I_g(\gamma_i) : 1 \le i \le l\}$ *find* $h \in G$ *such that* $I_h = I_g$. *We write* $h \in \text{IAP}_G(I_g)$.

Remark 1. For $g \in G$, $x, y \in Z(G)$, and $a \in \mathbb{N}$ we have

$$\text{DLP}(g, g^a) \subseteq \text{gDLP}(g, g^a) = \text{gDLP}(xg, yg^a).$$

Note that the solution of an instance (g, h) of the *gDLP* in a group G is unique modulo $|<g>/<g> \cap Z(G)|$. Naturally, the problem is trivial if $g \in Z(G)$.

A solution of the IAP is not necessarily unique. In fact we have for $h \in G$:

$$I_h = I_g \Leftrightarrow h \in g \cdot Z(G),$$

i.e. $\text{IAP}_G(I_g) = g \cdot Z(G)$.

We refer to a computational problem P_1 as *polynomial-time reducible* to a problem P_2 (i.e. $P_1 \Rightarrow_P P_2$) if there exists a deterministic polynomial-time algorithm that solves P_1 and that may use an (unspecified) algorithm for P_2 as a subroutine. We also consider probabilistic algorithms. In this paper a probabilistic polynomial-time algorithm always refers to a polynomial Las-Vegas-algorithm (corresponding to the complexity class ZPP). With respect to such algorithms we define a computational problem to be *efficiently solvable*, if there exists a probabilistic polynomial time algorithm that solves P. And in the same way as above, we define P_1 *probabilistic polynomial-time reducible* to P_2, i.e. $P_1 \Rightarrow_{ZPP} P_2$.

Recall that the multiplication in a semidirect product $G \times_\theta H$ of two groups G and H is defined via a morphism $\theta : H \to Aut(G)$ as

$$(g_1, h_1) \cdot (g_2, h_2) = (g_1 \theta_{h_1}(g_2), h_1 h_2).$$

3   Reducing MOR on $GL(n, q) \times_\theta \mathcal{H}$

In [7] Paeng et al. propose to use the semidirect product $SL(2, p) \times_\theta \mathbb{Z}_p$ as a group for an implementation of the MOR scheme. On this group MOR has been analysed and found insecure by Tobias and Paeng et al. in [8,11,13]. Tobias describes two attacks that enable an attacker to determine the plaintext message up to one unknown variable without compromising the secret key. The techniques presented in his work use two properties:

A_1 The special type of the homomorphism $\theta : \mathbb{Z}_p \to Inn(SL(2, p))$ and the fact that the center of $SL(2, p)$ is trivial allow an immediate reduction of MOR on $SL(2, p) \times_\theta \mathbb{Z}_p$ to MOR on $SL(2, p)$.

A_2 In a MOR cryptosystem using the group $SL(2, p)$, a ciphertext is a conjugate of the plaintext, and thus both texts have the same eigenvalues. This is enough information to calculate two linear equations in the entries of these matrices. The additional (and efficient) computation of an element of the centralizer of the enciphering matrix allows an adversary to obtain a third linear equation. These equations relate the four unknown entries of the plaintext matrix up to only one unknown variable.

In view of these attacks we propose to analyse MOR in a more general case for which the attacks described above do not work. Consider groups of the type

$$GL(n, q) \times_\theta \mathcal{H}$$

where $q = p^m$ is a prime power, $n \in \mathbb{N}$, $\mathcal{H}$ is any finite abelian group, and θ is an efficiently computable homomorphism of $\mathcal{H}$ into $Aut(GL(n, q))$. For our purpose, it is sufficient to assume that group operations in the semidirect product can be efficiently computed and the representation problem is efficiently solvable.

Every automorphism in $Aut(GL(n, q))$ is a composition of an inner automorphism, a central automorphism, a field automorphism and a contragredient transformation. For a detailed description of this automorphism group we refer the reader to [2,4].

By choosing a group for MOR in this way, we avoid an easy reduction as in A_1 and also the attack A_2.

3.1   The DLP in $Inn(GL(n, q))$

Before looking at MOR on the semidirect product, we analyse the DLP in $Inn(GL(n, q))$. This is done for two reasons: 1. The attack A_2 is less efficient with growing n, thus we should also consider MOR on $GL(n, q)$. 2. In order to conduct the analysis of MOR in Sec. 3.2 we need to look at the reducibility of the DLP in $Inn(GL(n, q))$.

Let G_1 and G_2 be two matrices that generate $GL(n, q)$, for details see [10].

Reduction 1 - DLP in $Inn(GL(n, q))$. Let (I_C, I_{C^a}) be an instance of the DLP in $Inn(GL(n, q))$. If we solve the IAP for I_C and I_{C^a}, we obtain the sets $C \cdot Z(GL(n, q)) = C \cdot \mathbb{F}_q^*$ and $C^a \cdot \mathbb{F}_q^*$ respectively. Let $V \in C \cdot \mathbb{F}_q^*$ and $W \in C^a \cdot \mathbb{F}_q^*$, then by Remark 1 we have

$$gDLP(V, W) = gDLP(C, C^a) = DLP(I_C, I_{C^a}).$$

Thus, efficiently solving the IAP in $GL(n, q)$ reduces the DLP in $Inn(GL(n, q))$ to the gDLP in $GL(n, q)$.

Proposition 1. *The* IAP *in* $GL(n, q)$ *is efficiently solvable.*

Proof. Let $I_C \in Inn(GL(n, q))$. From $\bar{G}_i := I_C(G_i)$ $(i = 1, 2)$ we derive a system of linear equations for the unknown variable $X \in GL(n, q)$

$$XG_1 - \bar{G}_1 X = 0_n \tag{1}$$
$$XG_2 - \bar{G}_2 X = 0_n \tag{2}$$

(which is satisfied by $X = C$). This system of $2n^2$ linear equations in n^2 variables with coefficients in $\mathbb{F}_q$ is efficiently solvable. Naturally, the zero matrix 0_n is a solution, but it is the only singular solution:

Let $A \neq 0_n$ be a simultaneous solution of (1) and (2). Then

$$AG_i = \bar{G}_i A \quad \text{and thus} \quad AG_i = CG_i C^{-1} A,$$

which implies

$$(C^{-1}A)G_i = G_i(C^{-1}A).$$

Since G_1 and G_2 generate $GL(n,q)$, the matrix $C^{-1}A$ centralizes $GL(n,q)$ and thus also the $\mathbb{F}_q$-vectorspace generated by $GL(n,q)$ which is $M_n(q)$. Therefore $C^{-1}A$ is of the form $c \cdot E_n$, $c \in \mathbb{F}_q$. Since $A \neq 0_n$, we have $C^{-1}A \in Z(GL(n,q))$, and thus $A \in C \cdot Z(GL(n,q)) = \text{IAP}(I_C)$. □

Corollary 1. DLP in $Inn(GL(n,q)) \Rightarrow_P$ gDLP in $GL(n,q)$.

Reduction 2 - gDLP in $GL(n,q)$. We will now discuss a further reduction of the gDLP using an eigenvalue attack.

Proposition 2. *If the Jordan canonical form of a matrix in $GL(n,q)$ and the corresponding transformation matrix are efficiently computable, then there exists $d \leq n$ such that*

$$\text{gDLP in } GL(n,q) \Rightarrow_P \text{DLP in } GL(n,q^d).$$

Proof. Let $V \in GL(n,q)$, $W \in \mathbb{F}_q^* \cdot <V>$ and $\lambda \in \mathbb{F}_{q^d}$ be an eigenvalue of V for some $d \leq n$. Consider the matrix

$$\hat{V} = \lambda^{-1} \cdot V \in GL(n,q^d). \tag{3}$$

If $\lambda_1, \ldots, \lambda_s$ are the eigenvalues of V, then $\lambda^{-1}\lambda_1, \ldots, \lambda^{-1}\lambda_s$ are the eigenvalues of $\hat{V}$; and one of these equals 1. Without loss of generality we assume $\lambda^{-1}\lambda_1 = 1$.

Let $J(\hat{V})$ be the Jordan canonical form of $\hat{V}$, and let T be the corresponding transformation matrix, i.e. $T\hat{V}T^{-1} = J(\hat{V})$. For $a \in \text{gDLP}_{GL(n,q^d)}(\hat{V},W)$ there exists $k_a \in \mathbb{F}_q^*$ such that

$$J(\hat{V})^a = (T\hat{V}T^{-1})^a = T\hat{V}^aT^{-1} = k_aTWT^{-1}.$$

In this equation, $J(\hat{V})^a$ as well as TWT^{-1} are upper triangular matrices. The elements on the diagonal of $J(\hat{V})^a$ are

$$1, (\lambda^{-1}\lambda_2)^a, \ldots, (\lambda^{-1}\lambda_s)^a,$$

where 1 is in position (i,i) for some $1 \leq i \leq n$, Let w be the entry in position (i,i) of TWT^{-1}. Then we have

$$k_a = w^{-1}. \tag{4}$$

Now we know k_a, and thus the problem of solving the gDLP for $\hat{V}$ and W is reduced to the DLP in $GL(n,q^d)$ for $\hat{V}$ and k_aW. □

Note that in order to compute the Jordan canonical form and the transformation matrix in Proposition 2 we would need to do computations in the field $\mathbb{F}_{q^t}$, where $t = lcm\{d_1, \ldots, d_s\}$ for $\lambda_i \in \mathbb{F}_{q^{d_i}}$, and $t \in O(n!)$. Hence, an algorithm using the straight-forward strategy in the proof of Proposition 2 would have exponential running time. In order to make the reduction efficient, we use the idea of the previous proof to modify an algorithm by Menezes-Wu, see [6]. Their algorithm actually yields a probabilistic polynomial-time reduction of the DLP in $GL(n,q)$ to the DLP in some small extension fields of $\mathbb{F}_q$ and is modified as follows:

Algorithm 1. (Reduction of the gDLP in $GL(n,q)$ to the DLP in $\mathbb{F}_{q^{m_i}}$, $1 \leq i \leq s$.) Changes to the original algorithm from [6] are *emphazised*.

Input: Matrices $V, W \in GL(n,q)$ with $V^a = zW$ for some $z \in \mathbb{F}_q^*$.
Output: $l \in \mathrm{gDLP}_{GL(n,q)}(V,W)$.

I. *Use Ben-Or's Las-Vegas algorithm (see [1]) to compute one eigenvalue $\lambda \in \mathbb{F}_{\hat{q}}$ of V.*

II. *Compute $\hat{V} = \lambda^{-1} V \in GL(n,\hat{q})$ (as in (3)).*
   Observe that now 1 is an eigenvalue of $\hat{V}$.

III. *Start the Menezes-Wu algorithm with $\hat{V}$ and W in $GL(n,\hat{q})$:*
   1. Use the Hessenberg algorithm to find the characteristic polynomial $p_{\hat{V}}(x)$ of $\hat{V}$.
   2. Find the factorization of $p_{\hat{V}}(x)$ over $\mathbb{F}_{\hat{q}}$ using Ben-Or's Las-Vegas algorithm (see [1]): $p_{\hat{V}}(x) = f_1^{e_1} f_2^{e_2} \cdots f_s^{e_s}$, where each f_i is an irreducible polynomial of degree d_i.
      Let the roots of f_i in $\mathbb{F}_{\hat{q}^{d_i}}$ be α_{ij}, $1 \leq j \leq d_i$. Note that we may conveniently represent the field $\mathbb{F}_{\hat{q}^{d_i}}$ as $\mathbb{F}_{\hat{q}}[x]/(f_i(x))$. In this representation, we simply have $\alpha_{i1} = x$ and $\alpha_{ij} = x^{\hat{q}^{j-1}} \bmod f_i(x)$ for $1 \leq j \leq d_i$.
      Since 1 is an eigenvalue of $\hat{V}$, one of the $f_i's$ is equal to $x-1$. Rearrange the factorization of $p_{\hat{V}}(x)$ such that $f_1 = x-1$. Then, $d_1 = 1$ and $\alpha_{11} = 1$.
   3. For i from 1 to s do the following:
      3.1 For $l = 1, \ldots, c, c+1$ compute $(\hat{V} - \alpha_{i1}I)^l$ and $r_l = rank(\hat{V} - \alpha_{i1}I)^l$, where c is the smallest positive integer such that $r_c = r_{c+1}$.
      3.2 Find an eigenvector μ_i corresponding to α_{i1} by solving $(\hat{V} - \alpha_{i1}I)y = 0$.
      3.3 Construct a matrix $Q_i \in GL(n, \hat{q}^{d_i})$ with first column μ_i.
      3.4 *If $i = 1$, compute $D_1 = Q_1^{-1} W Q_1$.*
         If $i \geq 2$, compute $D_i = Q_i^{-1} \hat{W} Q_i$.
      3.5 *If $i = 1$, then the (1,1) entry of D_1 is w (as in (4)).*
         Set $\hat{W} = w_1^{-1} W$ and $l \bmod ord(\alpha_{11}) = l \bmod 1 = 1$.
         If $i \geq 2$, then the (1,1) entry of D_i is α_{i1}^l. Obtain $l \bmod ord(\alpha_{i1})$ by solving the DLP in $\mathbb{F}_{\hat{q}^{d_i}}$ for the instance $(\alpha_{i1}, \alpha_{i1}^l)$.
   4. Let t be the maximum of the values c found in step 3.1. If $t > 1$ then the value $l \bmod p\{t\}$ is computed in this step, see [6].
   5. Compute $l \bmod ord(\hat{V})$ using the Chinese remainder theorem (where $ord(\hat{V}) = lcm\{ord(\alpha_{ij})\} \cdot p\{t\}$).

Remark 2. The running time of ALGORITHM 1 is probabilistic polynomial: The original algorithm of Menezes-Wu is a probabilistic polynomial-time algorithm (see [6]); the changes performed in ALGORITHM 1 do not alter the complexity.

Corollary 2. DLP in $Inn(GL(n,q)) \Rightarrow_{ZPP}$ DLP in $\mathbb{F}_{q^i}$, $i = 1, \ldots, n..$

3.2 Reduction of MOR on $GL(n, q) \times_\theta \mathcal{H}$

For $C \in GL(n, q)$ and $h \in \mathcal{H}$ let the pair

$$\left(I_{(C,h)}, \; I_{(C,h)^a}\right) \tag{5}$$

of inner automorphisms of $GL(n, q) \times_\theta \mathcal{H}$ be the public key for a MOR encryption in this group. Under the assumption that we are able to solve the DLP in small extensions of the finite field $\mathbb{F}_q$, we will succesfully compute an equivalent of the secret key a or other information which allows us to decipher any message

$$\left((M', s), \; I_{(C,h)^b}\right) \tag{6}$$

where

$$(M', s) = I_{(C,h)^{ab}}(M, s). \tag{7}$$

Since $\mathcal{H}$ is abelian, the second component s of the plaintext is sent in clear and the actual ciphertext is

$$M' = p(ab) \cdot \theta_{h^{ab}}(M) \cdot \theta_s(p(ab)^{-1}), \tag{8}$$

where $p(x) := \prod_{i=0}^{x-1} \theta_{h^i}(C)$. For $s = 1$ a MOR encryption on $GL(n, q) \times_\theta \mathcal{H}$ is immediately reduced to $GL(n, q)$. Hence, we assume that $s \neq 1$.

In this paper we will consider $\theta(\mathcal{H})$ as a subgroup of one of the four basic automorphism groups of $GL(n, q)$, i.e. for $h \in \mathcal{H}$

1. $\theta_h \in Inn(GL(n, q))$, or
2. $\theta_h \in Aut_C(GL(n, q))$, where $\theta_h(A) \in \mathbb{F}_q^* \cdot A$ for all $A \in GL(n, q)$, or
3. $\theta_h \in <ct>$, where $ct(A) = (A^{-1})^t$ for all $A \in GL(n, q)$, or
4. $\theta_h \in <f>$, where f is the generator of the field automorphisms.

Since every automorphism of $GL(n, q)$ is a product of basic automorphisms, it can be shown that the results in each of these cases will lead to the general case.

Theorem 1. *Let $\mathcal{H}$ be an abelian group and $\theta : \mathcal{H} \to Aut(GL(n,q))$ be a homomorphism, where $\theta(\mathcal{H})$ is a subgroup of one of the four basic automorphism groups of $GL(n, q)$. Then the MOR cryptosystem on the group $GL(n, q) \times_\theta \mathcal{H}$ succumbs to a ciphertext-only attack by an adversary, who is able to solve the DLP in small extension fields of $\mathbb{F}_q$, i.e. in $\mathbb{F}_{q^i}$, $i = 1, \ldots, n$.*

Proof. For the security analysis we will define two functions, which extract useful information from (5) and (6).

Definition 2. *Let $A \in GL(n, q)$. Given (5) and (6), and without knowing a or b, we are always able to compute the following values, where $x \in \{1, a, b\}$:*

1) Since $I_{(C,h)^x}(A, \ 1) = (I_{p(x)}(\theta_{h^x}(A)), \ 1)$, we define

$$\Psi_1(x, A) := I_{p(x)}(\theta_{h^x}(A)). \tag{9}$$

Note that for $A \in Z(GL(n, q))$ we have $\Psi_1(x, A) = \theta_{h^x}(A)$.

2) Since $I_{(C,h)^x}(A, \ s) = (p(x) \ \theta_{h^x}(A) \ \theta_s(p(x)^{-1}), \ s)$, we define

$$\Psi_2(x, A) := p(x) \ \theta_{h^x}(A) \ \theta_s(p(x)^{-1}). \tag{10}$$

Note that we also know

$$\Psi_2(ab, M) = M'. \tag{11}$$

We know look at each of the four cases seperately. The first case is discussed in detail and we give a summary of our approach in the other cases.

CASE 1: $\theta(\mathcal{H}) \leq Inn(GL(n, q))$. If θ_h is an inner automorphism, then there exists $H \in GL(n, q)$ such that

$$\theta_h = I_H.$$

For $x \in \{1, a, b\}$ we have (but are not able to compute)

$$p(x) = \prod_{i=0}^{x-1} \theta_{h^i}(C) = \prod_{i=0}^{x-1} I_{H^i}(C)$$
$$= C \ HCH^{-1} \ H^2CH^{-2} \ \ldots \ H^{x-1}CH^{-(x-1)} \tag{12}$$
$$= (CH)^x H^{-x}.$$

Step 1. Consider Ψ_1 and only the public key (5). From (12) and for $x \in \{1, a\}$ and any matrix $A \in GL(n, q)$ we derive

$$\Psi_1(x, A) = I_{p(x)}(\theta_{h^x}(A))$$
$$= (CH)^x H^{-x} \theta_{h^x}(A) H^x (CH)^{-x} \tag{13}$$
$$= (CH)^x H^{-x} I_{H^x}(A) H^x (CH)^{-x} = I_{(CH)^x}(A).$$

This yields the inner automorphisms I_{CH} and $I_{(CH)^a}$. By Corollary 2 we efficiently solve the DLP for I_{CH} and $I_{(CH)^a}$ in $Inn(GL(n, q))$ and derive

$$a' \in \mathrm{DLP}(I_{CH}, \ I_{(CH)^a}) \tag{14}$$

Note that this computation can be done before any message (6) has been sent.

Step 2. Now, consider the encrypted message (6). Since we know s, we are able to calculate $S \in \mathrm{IAP}(\theta_s)$ such that

$$\theta_s = I_S.$$

This is not necessarily the same generating element of θ_s, which is used for the encryption, but we will see that the matrix S is sufficient for our purpose. (The matrix, which is used for encryption, is some $\mathbb{F}_q^*$-multiple of S).

Since $\theta(\mathcal{H})$ is abelian, we have

$$I_{HS} = I_{SH}.$$

It follows that

$$[S, H] = SHS^{-1}H^{-1} = c \cdot E_n \in Z(GL(n,q)). \tag{15}$$

for some $c \in \mathbb{F}_q^*$. Since $\det[S, H] = 1$, c is even a n-th root of unity in $\mathbb{F}_q^*$.

Now consider Ψ_2. From (12) and (15) we compute for $x \in \{1, a\}$ and some $A \in GL(n,q)$

$$\begin{aligned}
\Psi_2(x, A) &= p(x)\, \theta_{h^x}(A)\, \theta_s(p(x)^{-1}) \tag{16}\\
&= (CH)^x H^{-x} \cdot H^x A H^{-x} \cdot SH^x (CH)^{-x} S^{-1}\\
&= (CH)^x \cdot Ac^x S \cdot (CH)^{-x} \cdot S^{-1}\\
&= I_{(CH)^x}(Ac^x S)S^{-1}.
\end{aligned}$$

Since we know S (except for a $\mathbb{F}_q^*$-multiple), A, and $I_{(CH)^x}$, we are able to extract c, c^a, $c^b \in \mathbb{F}_q^*$. From these we can efficiently compute

$$c^{ab} \in \mathbb{F}_q^*.$$

As in Step 1. (13) we efficiently compute $I_{(CH)^b}$. Using a' we derive

$$I_{(CH)^{ab}}.$$

The encrypted part of the message is of the form

$$M' = I_{(CH)^{ab}}(Mc^{ab}S)S^{-1}. \tag{17}$$

Since we know S (except for a $\mathbb{F}_q^*$-multiple), c^{ab}, and the inner automorphism $I_{(CH)^{ab}}$, we are able to compute the plaintext

$$M = I_{(CH)^{-ab}}(M' \cdot S) \cdot (c^{ab}S)^{-1}.$$

Note that knowing a' is equivalent to knowing the secret key a. In this case MOR on $GL(n,q) \times_\theta \mathcal{H}$ is insecure.

CASE 2: $\theta(\mathcal{H}) \le Aut_c(GL(n,q))$. There exists $c \in \mathbb{F}_q^*$ whence the encrypted part of the message is of the form

$$M' = I_{C^{ab}}(c \cdot M). \tag{18}$$

Using Ψ_1 and the invariance of the trace function under conjugation we compute the inner automorphisms I_C, I_{C^a} and I_{C^b}. Solving a DLP we derive $I_{C^{ab}}$. Thus we are able to recover an $\mathbb{F}_q^*$-multiple of the message M. In this case we gain sufficient information (i.e. the plaintext) such that MOR on $GL(n,q) \times_\theta \mathcal{H}$ is insecure.

CASE 3: $\theta(\mathcal{H}) \leq\, < ct >$. Now, the encrypted part of the message is of the form

$$M' = \begin{cases} I_{p(ab)}(M) & \text{, if } \theta_s = id \\ p(ab) \cdot M \cdot p(ab)^t & \text{, if } \theta_s = ct \end{cases} \tag{19}$$

In the first case use Ψ_1 to compute $I_{p(1)}$, $I_{p(a)}$, and $I_{p(b)}$. Solving a DLP in $Inn(GL(n, q))$ we derive $I_{p(ab)}$ and thus the message M. In the second case use Ψ_1 and Ψ_2 to compute $p(1)$, $p(a)$, and $p(b)$. Solving a DLP in $GL(n, q)$ we derive $p(ab)$ and thus the message M. In both cases we compute an equivalent to the secret key a. Hence, MOR on $GL(n, q) \times_\theta \mathcal{H}$ is insecure.

CASE 4: $\theta(\mathcal{H}) \leq\, < f >$. The encrypted part of the message is of the form

$$M' = p(m)^{k_{ab}} \cdot p(ab_m) \cdot \theta_h^{ab_m}(M) \cdot \theta_s((p(m)^{k_{ab}} \cdot p(ab_m))^{-1}). \tag{20}$$

Since we know s, we are able to calculate θ_s. From Ψ_1 we derive θ_h. Using Ψ_1 and the extended euklidian algorithm we compute $ab_m = a \cdot b \bmod m$, where $m = \log_p q$. We can also efficiently compute $p(m)$ and $p(ab_m)$. Using Ψ_2 we compute $I_{p(m)}, I_{p(m)^{k_a}}$, and $I_{p(m)^{k_b}}$. Solving a DLP in $Inn(GL(n, q))$ we derive $k_{ab} = \frac{ab - ab_m}{m}$ and thus the message M and an equivalent to the secret key a. Hence, MOR is insecure in this case.   □

Remark 3. Note that a similar analysis is possible with $SL(n, q)$ in a semidirect product with any finite abelian group.

4   Generic Security Analysis of MOR

In order to break the security of the MOR cryptosystem in the generic case, it is sufficient to be able to solve the DLP in a cyclic subgroup of an inner automorphism group of the group G. To the author's best knowledge, the only algorithms known, which can be applied to the DLP in an inner automorphism group, are generic algorithms which have exponential running time. In [7] it is argued that even if the discrete logarithm problem in the underlying group G is efficiently solvable, their cryptosystem is applicable to G. Thus, if we are able to efficiently reduce the DLP from $Inn(G)$ to G and there exist (at least) subexponential algorithms for solving the DLP in G, this advantage of the MOR construction is annuled.

In the following we will give some results on the general interrelation between the DLP in an inner automorphism group $Inn(G)$ and the DLP in G.

Theorem 2. *(DLP in an inner automorphism group, I) Let G be a group and $g \in G$. If a solution of the IAP for I_g in G is computable in polynomial time, then*

$$\text{DLP in } < I_g > \Leftrightarrow_P \text{ gDLP in } <g>.$$

Proof. $\Rightarrow$: Let (I_g, I_h) be an instance of the DLP in $< I_g >$. If we solve the IAP for both inner automorphisms we obtain $x \in g \cdot Z(G)$ and $y \in h \cdot Z(G)$. Note that $I_g = I_x$ and $I_h = I_y$. If we solve the gDLP for the instance (x, y) we obtain $a \in \mathbb{N}$ such that $x^a = y$. With Remark 1 it follows that $a \in \text{gDLP}(x, y) = \text{gDLP}(g, h) = \text{DLP}(I_g, I_h)$.

$\Leftarrow$: Let (g, h) be an instance of the gDLP in G. Lift this instance to $< I_g >$, i.e. (I_g, I_h). For a solution $a \in \text{DLP}(I_g, I_h)$ we have $(I_g)^a = I_h$ and it follows that $g^a \in h \cdot Z(G)$. Thus $a \in \text{gDLP}_{<g>}(g, h)$. $\qquad\square$

Thus the security of the MOR cryptosystem depends on the difficulty of the gDLP in G and the difficulty of calculating solutions for the IAP in $Inn(G)$. Certain conditions on the order of g allow a further reduction.

Theorem 3. *(DLP in an inner automorphism group, II) Let G be a group and $g \in G$ such that $\gcd(ord(g), |Z(G)|) = 1$. If a solution of the IAP for I_g in G is computable in polynomial time, then*

$$\text{DLP in } < I_g > \Leftrightarrow_P \text{ DLP in } <g> .$$

Proof. By theorem 2 we need to prove that the DLP and the gDLP in $<g>$ are equivalent.
1. Since the condition $\gcd(ord(g), |Z(G)|) = 1$ guarantees the uniqueness of the solution of the gDLP, we have DLP in $<g> \Rightarrow$ gDLP in $<g>$.
2. Let $|Z(G)| = m$ and (g, h) be an instance of the gDLP in $< g >$. Since $\gcd(ord(g), m) = 1$, a solution $a \in \text{gDLP}(g, h)$ is unique modulo $ord(g)$. We also have $a \in \text{DLP}(g^m, h^m)$ and a is a unique solution modulo $ord(g^m) = ord(g)$. It follows that $\text{DLP}(g^m, h^m) = \text{gDLP}(g, h)$. $\qquad\square$

Note that the theorem also holds if $\gcd(ord(g), | Z(G) |)$ is small, i.e. in $O(\log ord(g))$. This theorem implies that in certain cases it is indeed important for the security of MOR whether the discrete logarithm problem in the underlying group is efficiently solvable. Especially groups, where the order of the center and its factor group are coprime, must not be used.

5   Conclusion

We showed that MOR on groups of the type $GL(n, q) \times_\theta \mathcal{H}$ is insecure in the sense that an attacker, who is able to compute discrete logarithms in small extensions of $\mathbb{F}_q$, is also able to break MOR. As the DLP in finite fields offers high security, the semidirect products considered in this paper should be discarded for practical reasons: computations in these are more costly than in the corresponding finite fields $\mathbb{F}_{q^i}$. Thus an ElGamal-type cryptosystem in $\mathbb{F}_{q^i}$ is more efficent with the same security. The question, whether there exist groups on which MOR is secure, remains open. When studying MOR on other groups the considerations of the previous section must be taken into account.

References

1. Ben-Or, M.: Probabilistic algorithms in finite fields. IEEE Symposium on Foundations of Computer Science 22, 394–398 (1981)
2. Dieudonné, J.: On the automorphisms of classical groups. Memoirs of the American Mathematical Society 2 (1951)

3. ElGamal, T.: A Public-Key Cryptosystem and a Signature Scheme Based on Discrete Logarithms. In: Blakely, G.R., Chaum, D. (eds.) CRYPTO 1984. LNCS, vol. 196. pp. 10–18. Springer, Heidelberg (1985)
4. Korsten, A.: The Discrete Logarithm Problem in Semidirect Products and the Reduction of the MOR Cryptosystem, Diplomarbeit, University of Tübingen, Germany (2005)
5. Lee, I.S., Kim, W.H., Kwon, D., Nahm, S., Kwak, N.S., Baek, Y.J.: On the security of MOR public key cryptosystem. In: Lee, P.J. (ed.) ASIACRYPT 2004. LNCS, vol. 3329. pp. 387–400. Springer, Heidelberg (2004)
6. Menezes, A.J., Wu, Y.: The discrete logarithm problem in GL(n,p). Ars Combinatoria 47, 23–32 (1998)
7. Paeng, S.H., Ha, K.C., Kim, J.H., Chee, S., Park, C.: New public key cryptosystem using finite non abelian groups. In: Kilian, J. (ed.) CRYPTO 2001. LNCS, vol. 2139. pp. 470–485. Springer, Heidelberg (2001)
8. Paeng, S.H., Kwon, D., Ha, K.C., Kim, J.H.: Improved public key cryptosystem using finite non abelian groups, Cryptology ePrint Archive (2001), http://eprint.iacr.org/2001/066
9. Paeng, S.H.: On the security of cryptosystem using automorphism groups. Information Processing Letters 88, 293–298 (2003)
10. Taylor, D.: Pairs of generators for matrix groups, I. The Cayley Bulletin 3, 76–85 (1987)
11. Tobias, C.: Security analysis of the MOR cryptosystem. In: Desmedt, Y.G. (ed.) PKC 2003. LNCS, vol. 2567. pp. 175–186. Springer, Heidelberg (2002)
12. Tobias, C.: Security analysis of MOR using $GL(2, R) \times \mathbb{Z}_p$. WOSIS 2, 170–179 (2004)
13. Tobias, C.: Design und Analyse kryptographischer Bausteine auf nicht-abelschen Gruppen, PhD thesis, University of Giessen (2004)

Preimages for Reduced-Round Tiger[*]

Sebastiaan Indesteege[**] and Bart Preneel

Katholieke Universiteit Leuven, Dept. ESAT/SCD-COSIC,
Kasteelpark Arenberg 10, B-3001 Heverlee, Belgium
{sebastiaan.indesteege,bart.preneel}@esat.kuleuven.be

Abstract. The cryptanalysis of the cryptographic hash function Tiger has, until now, focussed on finding collisions. In this paper we describe a preimage attack on the compression function of Tiger-12, i.e., Tiger reduced to 12 rounds out of 24, with a complexity of $2^{63.5}$ compression function evaluations. We show how this can be used to construct second preimages with complexity $2^{63.5}$ and first preimages with complexity $2^{64.5}$ for Tiger-12. These attacks can also be extended to Tiger-13 at the expense of an additional factor of 2^{64} in complexity.

Keywords: Tiger, hash functions, preimages.

1   Introduction

A cryptographic hash function is expected to possess three properties: collision resistance, preimage resistance and second preimage resistance. While other properties exist, the above three are the most well known.

Collision resistance: It is difficult to find two distinct messages $m \neq m'$ that hash to the same result, i.e., $h(m) = h(m')$.

Preimage resistance: When given a hash result y (for which it holds that $\exists x : h(x) = y$), it is difficult to find a message m which hashes to y, i.e., $h(m) = y$.

Second preimage resistance: When given a message m, it is difficult to find a message $m' \neq m$ that hashes to the same result as the given message, i.e., $h(m) = h(m')$.

There are generic attacks that apply to any hash function and whose time complexity only depends on the size of the hash result. Collisions for a hash function with an n-bit result can be found in time $2^{n/2}$ using a birthday attack [6], and preimages can be found by brute force in time 2^n. Weaker attacks may aim

---

[*] This work was supported in part by the Concerted Research Action (GOA) Ambiorics 2005/11 of the Flemish Government, by the IAP Programme P6/26 BCRYPT of the Belgian State (Belgian Science Policy), and by the European Commission through the IST Programme under Contract IST-2002-507932 ECRYPT.

[**] F.W.O. Research Assistant, Fund for Scientific Research – Flanders (Belgium).

S. Lucks, A.-R. Sadeghi, and C. Wolf (Eds.): WEWoRC 2007, LNCS 4945, pp. 90–99, 2008.

at finding pseudo-collisions, where slight differences in the hash results are allowed, or pseudo-near-collisions, where differences may also appear in the initial chaining values.

All attacks on the cryptographic hash function Tiger [1] have so far been collision attacks. Kelsey and Lucks [3] showed a collision attack on Tiger reduced to 16 rounds with a complexity of 2^{44} compression function evaluations. Mendel et al. [4] extended this to a collision attack on 19 rounds of Tiger with a complexity of 2^{62} compression function evaluations. In both papers some weaker attacks (e.g. pseudo-collisions) for a larger number of rounds were also shown. These results were further improved by Mendel et al. [5] towards a pseudo-near-collision for the full hash function and a pseudo-collision for 23 rounds of Tiger.

We focus on finding preimages for reduced variants of Tiger instead. More specifically, we describe a method to find first and second preimages for 12 and 13 rounds reduced Tiger. This method is conceptually similar to Dobbertin's preimage attack on reduced MD4 [2]. Our attack finds first and second preimages for Tiger-12 with a complexity of $2^{64.5}$ and $2^{63.5}$ compression function evaluations, respectively. It can be extended to Tiger-13, where the complexities become $2^{128.5}$ and $2^{127.5}$, respectively. As Tiger has a digest size of 192 bits, the theoretical complexity for finding first or second preimages is 2^{192} compression function evaluations. To the best of our knowledge, this is the first result concerning preimages for reduced round Tiger.

The structure of the paper is as follows. In Sect. 2, the Tiger hash function is described, along with the notation that will be used throughout the paper. Section 3 describes a preimage attack on three rounds of Tiger. The three round preimage attack is then used as a building block to construct preimages for the compression function of Tiger-12 and Tiger-13 in Sect. 4. Then, in Sect. 5 it is shown how first and second preimages for these reduced variants of the Tiger hash function can be constructed. Finally, Sect. 6 presents our conclusions.

2   Description of Tiger

Tiger [1] is an iterative cryptographic hash function, designed by Anderson and Biham in 1996. It has an output size of 192 bits. Truncated variants with a digest size of 160 and 128 bits were also defined. It was designed for 64-bit architectures and hence all words are 64 bits wide and arithmetic is performed modulo 2^{64}. Tiger uses the little-endian byte ordering.

First, the message to be hashed is padded by appending a single "1"-bit and as many "0"-bits as necessary to make the message length 64 bits less than the next multiple of 512 bits. Then the message length (in bits) is appended as a 64-bit unsigned integer. After this procedure, the padded message consists of an integer number of 512-bit blocks. Then, Tiger's compression function is applied iteratively to each 512-bit block of the padded message.

Tiger's compression function operates on a 192-bit chaining value and a 512-bit message block. The message block is split into eight 64-bit words X_i. The 192-bit chaining value is split into three 64-bit words which are used as the

Table 1. Notations

$X + Y$	Addition of X and Y modulo 2^{64}
$X - Y$	Subtraction of X and Y modulo 2^{64}
$X \times Y$	Multiplication of X and Y modulo 2^{64}
$X \oplus Y$	Bit-wise exclusive or of X and Y
$\overline{X}$	Bit-wise complement of X
$X \ll n$	Logical left bit shift of X by n positions
$X \gg n$	Logical right bit shift of X by n positions
$X\|Y$	The concatenation of X and Y
X_i	The i-th expanded message word
Y_i	The i-th intermediate value of the key schedule algorithm
A_i, B_i, C_i	State variables at the output of round i, $0 \leq i < 24$
K_i	The round constant used in round i, $0 \leq i < 24$
K_i^{-1}	Multiplicative inverse of K_i modulo 2^{64}
$T_1,\ldots,T_4$	The four 8-to-64-bit S-boxes used in Tiger

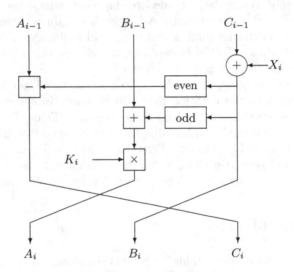

Fig. 1. The state update transformation of Tiger

initial state variables A_{-1}, B_{-1} and C_{-1}. The compression function consists of three passes of 8 rounds of a state update transformation (24 rounds in total), each using one X_i to update the three state variables A_i, B_i and C_i. Table 1 summarises the notations used in this paper.

The i-th round of Tiger ($0 \leq i < 24$) is depicted in Fig. 1. Equivalently, the state update transformation can be described by the following equations:

$$
\begin{aligned}
A_i &= K_i \times (B_{i-1} + \mathrm{odd}\,(C_{i-1} \oplus X_i)) \ , \\
B_i &= C_{i-1} \oplus X_i \ , \\
C_i &= A_{i-1} - \mathrm{even}\,(C_{i-1} \oplus X_i) \ .
\end{aligned}
\tag{1}
$$

In every round, a round constant K_i is used. These constants are given by:

$$K_i = \begin{cases} 5 & \text{if } 0 \le i < 8 , \\ 7 & \text{if } 8 \le i < 16 , \\ 9 & \text{if } 16 \le i < 24 . \end{cases} \tag{2}$$

The non-linear functions odd$(\cdot)$ and even$(\cdot)$ are defined as follows.

$$\begin{aligned} \text{odd}(c_7||\dots||c_0) &= T_4[c_1] \oplus T_3[c_3] \oplus T_2[c_5] \oplus T_1[c_7] , \\ \text{even}(c_7||\dots||c_0) &= T_1[c_0] \oplus T_2[c_2] \oplus T_3[c_4] \oplus T_4[c_6] . \end{aligned} \tag{3}$$

Here, c_i denotes the i-th byte of a 64-bit word, using the little-endian byte ordering, i.e., c_0 is the least significant byte.[1] Both functions use four 8-to-64-bit S-boxes, T_1 through T_4. Note that both functions only use four out of eight input bytes, and thus map 32 bits to 64 bits. They are called odd$(\cdot)$ and even$(\cdot)$ because they operate on the odd, respectively even bytes of the input word.

The first eight message words $X_i, 0 \le i < 8$, are taken directly from the message block. The message words $X_8,\dots,X_{15}$ are derived from $X_0,\dots,X_7$ using an algorithm which the designers of Tiger refer to as the key schedule algorithm [1]. Then, using the same algorithm, $X_{16},\dots,X_{23}$ are determined from $X_8,\dots,X_{15}$. This key schedule algorithm consists of two passes, given by the following equations:

$$\begin{aligned} Y_0 &= X_0 - (X_7 \oplus \text{A5}\dots\text{A5}_x) , & X_8 &= Y_0 + Y_7 , \\ Y_1 &= X_1 \oplus Y_0 , & X_9 &= Y_1 - \left(X_8 \oplus (\overline{Y_7} \ll 19)\right) , \\ Y_2 &= X_2 + Y_1 , & X_{10} &= Y_2 \oplus X_9 , \\ Y_3 &= X_3 - \left(Y_2 \oplus (\overline{Y_1} \ll 19)\right) , & X_{11} &= Y_3 + X_{10} , \\ Y_4 &= X_4 \oplus Y_3 , & X_{12} &= Y_4 - \left(X_{11} \oplus (\overline{X_{10}} \gg 23)\right) , \\ Y_5 &= X_5 + Y_4 , & X_{13} &= Y_5 \oplus X_{12} , \\ Y_6 &= X_6 - \left(Y_5 \oplus (\overline{Y_4} \gg 23)\right) , & X_{14} &= Y_6 + X_{13} , \\ Y_7 &= X_7 \oplus Y_6 . & X_{15} &= Y_7 - (X_{14} \oplus \text{01}\dots\text{EF}_x) . \end{aligned} \tag{4}$$

Finally, after 24 rounds, the initial state variables are fed forward, using a combination of exclusive or, subtraction and addition.

$$\begin{aligned} A^\star &= A_{-1} \oplus A_{23} , \\ B^\star &= B_{-1} - B_{23} , \\ C^\star &= C_{-1} + C_{23} . \end{aligned} \tag{5}$$

The 192-bit output of the compression function is $A^\star||B^\star||C^\star$, i.e., the concatenation of $A^\star$, $B^\star$ and $C^\star$.

3   Preimages for Three Rounds of Tiger

In this section we describe a solution due to Mendel et al. [4] to the problem of finding preimages for three rounds of the state update transformation of Tiger.

---

[1] Note that there was a misinterpretation of the byte order in [3,4]. The attacks described there can however be modified to overcome this problem [5].

There is always exactly one solution, which can be found in constant time. Although rather straightforward, it will prove to be a useful building block in preimage attacks on a larger number of Tiger rounds.

More in detail, we are given A_{-1}, B_{-1}, C_{-1}, A_2, B_2 and C_2 and want to determine the three message words X_0, X_1 and X_2 such that the constraints originating from the state update transformation are satisfied. Note that, without knowing any of the message words, all the state variables in these three rounds can already be determined. Indeed, from (1) it follows that

$$\begin{aligned}
A_1 &= C_2 + \text{even}\,(B_2) \ , \\
B_1 &= \left(A_2 \times K_2^{-1}\right) - \text{odd}\,(B_2) \ , \\
B_0 &= \left(A_1 \times K_1^{-1}\right) - \text{odd}\,(B_1) \ , \\
A_0 &= K_0 \times (B_{-1} + \text{odd}\,(B_0)) \ , \\
C_0 &= A_{-1} - \text{even}\,(B_0) \ , \\
C_1 &= A_0 - \text{even}\,(B_1) \ .
\end{aligned}$$

(6)

Note that each K_i as given in (2) is coprime with 2^{64} so its multiplicative inverse modulo 2^{64} exists and can be computed easily. Knowing the state variables, it is trivial to determine X_0, X_1 and X_2.

$$\begin{aligned}
X_0 &= C_{-1} \oplus B_0 \ , \\
X_1 &= C_0 \oplus B_1 \ , \\
X_2 &= C_1 \oplus B_2 \ .
\end{aligned}$$

(7)

This procedure is fully deterministic and always gives exactly one solution. The time complexity of this procedure is equivalent to three rounds of Tiger.

Of course this can equally be applied to any three consecutive rounds of Tiger, as part of a larger attack. To conclude, control over three consecutive expanded message words yields complete control over the intermediate state of Tiger.

4   Preimages for the Compression Function of Tiger-12

In this section, we first describe a method to find preimages for the compression function of Tiger, reduced to 12 rounds. Then we extend this to Tiger-13, i.e., Tiger reduced to 13 rounds.

Given the algorithm from Sect. 3, one can easily find sets of expanded message words X_i which ensure that the output of the compression function of Tiger (or a round-reduced version thereof) is equal to some desired value. However, if the number of attacked rounds is greater than eight there is no guarantee that these expanded message words satisfy the constraints from the key schedule algorithm. For eight or less rounds of Tiger, the message expansion becomes trivial, as each of the first eight expanded message words is under direct control of an adversary. Hence also finding preimages for these variants of Tiger is trivial by making arbitrary choices and using the algorithm from Sect. 3 for the last three rounds.

The circular dependency can be broken by guessing some intermediate variable(s) and later verifying if the guess was correct. If the guess was wrong, the attack is simply repeated. Hence the time complexity of the attack is highly dependent on the probability that the correct guess was made. Since we assume that every value for the guessed variables is equally likely, this probability is equal to 2^{-n} where n is the total number of guessed bits.

Conceptually, this approach is very similar to the work of Dobbertin [2] on finding preimages for a reduced variant of MD4. Of course the similarity only exists on a very high level, due to the fact that MD4 and Tiger are very different hash functions.

4.1   Algorithm

In this section, a detailed description of the algorithm for finding preimages for the compression function of Tiger-12 is given. As we are given the desired input and output chaining values, the feed-forward given in (5) can easily be removed. Therefore, the state variables A_{-1}, B_{-1}, C_{-1}, A_{11}, B_{11} and C_{11} are known at the beginning of the attack.

1. Make arbitrary choices for the message words used in the four last rounds, (i.e. X_8, X_9, X_{10} and X_{11}). The state update transformation can be used in the backwards direction to determine A_7, B_7 and C_7, as follows:

$$\begin{cases} A_{i-1} = C_i + \text{even}\,(B_i) \ , \\ B_{i-1} = \left(A_i \times K_i^{-1}\right) - \text{odd}\,(B_i) \ , & \text{for } i = 11, \ldots, 8 \\ C_{i-1} = B_i \oplus X_i \ . \end{cases} \tag{8}$$

2. Guess Y_7, an intermediate value of the key schedule algorithm. This 64-bit guess is the only guess that will be made in the attack. It will be verified in the final step of the attack.

3. The message words X_8 through X_{11} are normally computed from the key schedule. These equations can easily be inverted to find the intermediate values Y_0, Y_1, Y_2 and Y_3 for which the values chosen in step 1 will appear:

$$\begin{aligned} Y_0 &= X_8 - Y_7 \ , \\ Y_1 &= X_9 + \left(X_8 \oplus \left(\overline{Y_7} \ll 19\right)\right) \ , \\ Y_2 &= X_{10} \oplus X_9 \ , \\ Y_3 &= X_{11} - X_{10} \ . \end{aligned} \tag{9}$$

This step is deterministic and always leads to a single solution. Looking further at the key schedule, the message words X_1 through X_3 can also be determined uniquely:

$$\begin{aligned} X_1 &= Y_1 \oplus Y_0 \ , \\ X_2 &= Y_2 - Y_1 \ , \\ X_3 &= Y_3 + \left(Y_2 \oplus \left(\overline{Y_1} \ll 19\right)\right) \ . \end{aligned} \tag{10}$$

4. Choose X_7 (there are 2^{64} choices) and compute X_0 using the key schedule:

$$X_0 = Y_0 + (X_7 \oplus \texttt{A5A5A5A5A5A5A5A5}_x) \tag{11}$$

5. Now, the first four expanded message words (i.e. X_0 through X_3) are known. The state update transformation can thus be used in the forward direction to calculate A_3, B_3 and C_3.

$$\begin{cases} A_i = K_i \times (B_{i-1} + \mathrm{odd}\,(C_{i-1} \oplus X_i)) \ , \\ B_i = C_{i-1} \oplus X_i \ , \\ C_i = A_{i-1} - \mathrm{even}\,(C_{i-1} \oplus X_i) \ . \end{cases} \qquad \text{for } i = 0, \ldots, 3 \qquad (12)$$

Similarly, as X_7 is known, the state update transformation can be applied in the backwards direction to calculate A_6, B_6 and C_6.

$$\begin{aligned} A_6 &= C_7 + \mathrm{even}\,(B_7) \ , \\ B_6 &= \left(A_7 \times K_7^{-1}\right) - \mathrm{odd}\,(B_7) \ , \\ C_6 &= B_7 \oplus X_7 \ . \end{aligned} \qquad (13)$$

6. Note that, because A_3, B_3, C_3, A_6, B_6 and C_6 are now known, the algorithm from Sect. 3 can be applied to determine the unique solution for X_4, X_5 and X_6.

$$\begin{aligned} A_5 &= C_6 + \mathrm{even}\,(B_6) \ , \\ B_5 &= \left(A_6 \times K_6^{-1}\right) - \mathrm{odd}\,(B_6) \ , \\ B_4 &= \left(A_5 \times K_5^{-1}\right) - \mathrm{odd}\,(B_5) \ , \\ A_4 &= K_4 \times (B_3 + \mathrm{odd}\,(B_4)) \ , \\ C_4 &= A_3 - \mathrm{even}\,(B_4) \ , \\ C_5 &= A_4 - \mathrm{even}\,(B_5) \ , \\ X_4 &= C_3 \oplus B_4 \ , \\ X_5 &= C_4 \oplus B_5 \ , \\ X_6 &= C_5 \oplus B_6 \ . \end{aligned} \qquad (14)$$

7. Finally, apply the key schedule, which is given in (4), to compute the correct value for Y_7 from the message words X_0 through X_7, all of which have now been determined. Verify if the guess for Y_7 made in step 2 is correct. If it is, a preimage has been found. If not, restart from step 4 with a different choice for X_7.

The probability that the guess for Y_7 is correct is 2^{-64} so we expect to find a preimage after 2^{64} tries. Note that one attempt requires just 8 rounds of the state update transformation and 5 equations of the key schedule algorithm, which is only about 2/3 of the computations of a compression function evaluation. For simplicity, we assume that every equation of the key schedule algorithm takes an equivalent amount of work. Hence, the overall complexity of the attack is equivalent to slightly less than $2^{63.5}$ evaluations of the compression function.

4.2   Extension to Tiger-13

The attack can be extended to 13 rounds, by additionally guessing the value of X_{12} before the attack and verifying if the guess was correct afterwards. This again happens with a probability of 2^{-64}, yielding a total complexity of $2^{127.5}$. While it is theoretically possible to make an extension towards 14 rounds of Tiger, this hardly has an advantage over a simple exhaustive search.

5   First and Second Preimages for Tiger-12

The technique that has been developed in the previous section will now be applied to construct first and second preimages for Tiger-12. An extension of this construction to Tiger-13 is also possible.

5.1   Second Preimages for Tiger-12

Figure 2 shows how second preimages for Tiger-12 can be constructed, for (padded) messages with at least two message blocks and no padding bits in the first message block. This is equivalent to the requirement that the given message is at least 512 bits long.

In order to circumvent any issues that arise from the padding (which includes the message length) we choose the length of the preimage to be equal to that of the given message. We can hence reuse the last message block from the given message. All message blocks from the beginning up to the second to last message block can be chosen arbitrarily. This leaves us with exactly one message block, the central block in Fig. 2. Because the chaining values are known before and after this block, the attack from Sect. 4 can be applied. Of course a trivial generalisation where more than one message block is copied from the given message exists. In this case, the attack is applied to an earlier message block instead.

This procedure to find second preimages adds negligible overhead to the attack as described in Sect. 4. Hence, the time complexity remains at $2^{63.5}$ evaluations of the Tiger-12 compression function.

5.2   First Preimages for Tiger-12

Finding first preimages is a bit more involved due to the fact that there is no given message which can be used to easily circumvent issues originating from the padding. To construct first preimages for Tiger-12, we proceed as follows.

First we choose the message length such that only a single bit of padding will be placed in X_6 of the last message block. This is equivalent to choosing a message length of $k \cdot 512 + 447$ bits, where k is a positive integer. Next, as shown

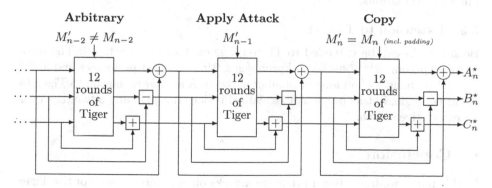

Fig. 2. Constructing second preimages for Tiger-12

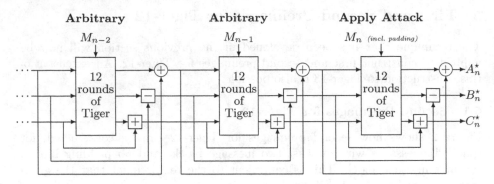

Fig. 3. Constructing first preimages for Tiger-12

in Fig. 3, all message blocks besides the last one are chosen arbitrarily and the attack is applied to this last block.

By choosing the message length in this way, X_7 of the last message block contains the message length as a 64-bit integer, which is fixed. Hence we can no longer choose X_7 freely during step 4 of the attack. By using the freedom in the choice of Y_7 in step 2 instead, the attack still works. Because step 3 is now also repeated, a larger part of the key schedule has to be redone on every attempt. The complexity figure of $2^{63.5}$ compression function evaluations can however be maintained because even with the larger part of the key schedule, the work of a single attempt does not exceed 70% — a fraction $2^{-0.5}$ — of a compression function evaluation. But additionally, we have to verify if the last bit of X_6 is a "1", as dictated by the padding rule. This happens with probability 2^{-1}, resulting in an overall complexity of $2^{64.5}$ compression function evaluations.

Note that the first preimages constructed in this way do not contain an integer number of bytes, which may not be acceptable. This problem can be solved by choosing the message length equal to $k \cdot 512 + 440$ bits instead. The only difference is that X_6 of the last message block now contains an entire byte of padding. The probability that this byte turns out to be correct after executing the attack is only 2^{-8}, and hence the overall complexity increases to $2^{71.5}$ compression function evaluations.

5.3 Extension to Tiger-13

Both attacks can be extended to Tiger-13, as explained in Sect. 4.2. The complexities become $2^{127.5}$ for second preimages, $2^{128.5}$ for first preimages and $2^{135.5}$ for first preimages of an integer number of bytes. A similar extension to Tiger-14 could be made, but as previously explained it does not give any advantage over an exhaustive search.

6   Conclusion

In this paper we have shown preimage attacks on reduced variants of the Tiger hash function. A method to find preimages for the compression function of

Tiger-12 and Tiger-13 with a complexity of $2^{63.5}$ and $2^{127.5}$, respectively, was described. It was shown how to construct first and second preimages for these variants of Tiger based on this method. To the best of our knowledge, this is the first result with respect to preimages of the Tiger hash function.

Acknowledgements

We would like to thank Florian Mendel, Christian Rechberger, Hirotaka Yoshida for interesting discussions, and the anonymous reviewers for their helpful comments.

References

1. Anderson, R., Biham, E.: Tiger: A Fast New Hash Function. In: Gollmann, D. (ed.) FSE 1996. LNCS, vol. 1039. pp. 89–97. Springer, Heidelberg (1996)
2. Dobbertin, H.: The First Two Rounds of MD4 are Not One-Way. In: Vaudenay, S. (ed.) FSE 1998. LNCS, vol. 1372. pp. 284–292. Springer, Heidelberg (1998)
3. Kelsey, J., Lucks, S.: Collisions and Near-Collisions for Reduced-Round Tiger. In: Robshaw, M. (ed.) FSE 2006. LNCS, vol. 4047. pp. 111–125. Springer, Heidelberg (2006)
4. Mendel, F., Preneel, B., Rijmen, V., Yoshida, H., Watanabe, D.: Update on Tiger. In: Barua, R., Lange, T. (eds.) INDOCRYPT 2006. LNCS, vol. 4329. pp. 63–79. Springer, Heidelberg (2006)
5. Mendel, F., Rijmen, V.: Cryptanalysis of the Tiger Hash Function. In: Kurosawa, K. (ed.) ASIACRYPT 2007. LNCS, vol. 4833. pp. 536–550. Springer, Heidelberg (2007)
6. Preneel, B.: Cryptographic primitives for information authentication – state of the art. In: Preneel, B., Rijmen, V. (eds.) State of the Art in Applied Cryptography. LNCS, vol. 1528. pp. 50–105. Springer, Heidelberg (1998)

Specific S-Box Criteria in Algebraic Attacks on Block Ciphers with Several Known Plaintexts

Nicolas T. Courtois[1] and Blandine Debraize[2,3]

[1] University College of London, Gower Street, London, UK
[2] Gemalto, Meudon, France
[3] University of Versailles, France

Abstract. In this paper we study algebraic attacks on block ciphers that exploit several (i.e. more than 2) plaintext-ciphertext pairs. We show that this considerably lowers the maximum degree of polynomials that appear in the attack, which allows much faster attacks, some of which can actually be handled experimentally. We point out a theoretical reason why such attacks are more efficient, lying in certain types of multivariate equations that do exist for some S-boxes. Then we show that when the S-box is on 3 bits, such equations do always exist. For S-boxes on 4 bits, the existence of these equations is no longer systematic. We apply our attacks to a toy version of Serpent, a toy version of Rijndael, and a reduced round version of Present, a recently proposed lightweight block cipher. It turns out that some S-boxes are much stronger than others against our attack.

Keywords: algebraic attacks on block ciphers, Rijndael, Serpent, multivariate equations, Gröbner bases, design of S-boxes, algebraic immunity.

1 Introduction

Algebraic attacks on block ciphers [5, 6] study different ways of describing the problem of recovering the secret key as a system of multivariate equations, especially those that make these systems efficiently solvable. In typical block ciphers such as AES or DES, one or two plaintext-ciphertext pairs give enough information to uniquely describe the key. Thus most cryptanalytic work done on this topic only considered systems of equations produced by only one or two plaintext-ciphertext pairs (see [6, 15, 16, 17, 18, 19]). Algebraic attacks have only been proven powerful on some very special ciphers like in [3, 4, 8, 18]. But very little is known about the resistance of the other block ciphers to algebraic attacks.

We will call here a plaintext-ciphertext pair, a **P-C Pair**. In this paper we show that using several plaintext-ciphertext pairs can considerably decrease the complexity of an algebraic attack. It will allow to keep all the polynomials at very low degree during the computation of Gröbner bases algorithm. We consider mostly the cases where the S-Boxes are very small (on 3 or 4 bits) and show that there is a specific reason why such small S-boxes are more vulnerable against algebraic attacks. For the toy cipher CTC proposed in [8] we are able to explain

S. Lucks, A.-R. Sadeghi, and C. Wolf (Eds.): WEWoRC 2007, LNCS 4945, pp. 100–113, 2008.

the reason why it is possible to break as many as 6 rounds of it. We also apply our attack methodology to reduced-round versions of Present, a new RFID-oriented block cipher proposed in [20]. We will see that the resistance of a few rounds of Present against algebraic attacks is not as good as announced by the authors.

This paper is organised as follows: in Section 2 we will briefly describe the theory of algebraic attacks, in Section 3 we describe our toy ciphers and analyse the computations we made on them, and in Section 4 we propose some cryptanalytic attacks that exploit the vulnerabilities of small S-boxes considered. Finally, in Section 5 we prove that very small S-boxes are always vulnerable to such attacks, and derive a new security criterion for S-boxes. Finally, Section 6 concludes the paper.

2   Preliminaries – Algebraic Attacks

Two stages are necessary to perform an algebraic attack on a block cipher:

2.1   Writing the Equations

The way one must write the system of equations completely depends on the algorithm used at the second stage of the attack. For algorithms belonging to the Gröbner Bases family, it is well known that the more overdefined is a system is, the more efficient will be the attack, see [2].

In this paper we use a very straightforward simple method for writing the equations of our toy cipher: the linear diffusion and key injection parts are described by the very equations that define it, and the S-boxes are defined by a basis of a system of all implicit quadratic equations that exist for this S-box.

2.2   Solving the Equations

One of the most powerful tool to solve this system of algebraic equations are linear algebra based algorithms like XL ([7]) or Gröbner basis algorithm like F4 or F5 ([11, 12]). In a nutshell (more details below), XL is an algorithm which aim is to solve a system of algebraic equations, while F4 computes a Gröbner basis of the ideal defined by the set of equations. We define a Gröbner basis of an ideal below. Both algorithms extend the idea of Gaussian elimination of systems of linear equations. All the algorithms such as XL, F4, F5 (and other) have two main distinct steps, that can be summarized as follows:

- **Expansion[D]:** This stage consists in multiplying some polynomials by some monomials. The maximum degree of the obtained polynomials is D.
- **Linearization:** Each different monomial in the equations is considered as a new variable. Then the linear system is solved by Gaussian elimination.

The exact method how these stages are computed depends on the algorithm and its implementation. However in all the three cases (XL, F4, F5) the complexity of the attack essentially depends on the complexity of the gaussian elimination when D is maximal : $\mathcal{T}^w$, where $\mathcal{T}$ is the number of variables of the linear system, and w depends on the chosen algorithm. We choose here $w = 3$, corresponding to the most practical algorithm. After the Expansion step, the number of variables T of the system we consider is then the number of monomials appearing in all the polynomials of the system. Thus if n is the number of variables of the algebraic system and D the maximal degree of the equations occurring during the computation, an theoretical estimation of this complexity is $\mathcal{O}(\binom{n}{D}^3)$. In practice, the complexity may decreases with the sparsity of the equations.

To perform the Gaussian elimination, we need an ordering $<$ on the monomials. XL algorithm only needs a total order, while Gröbner bases algorithms require a more precise type of ordering. We give some useful definitions below.

Let k be a field and $k[x_1, ..., x_n]$ its ring of polynomials in n variables. We use the following notation to represent a monomial: if $a = (a_1, ..., a_n)$, we write x^a for $x_1^{a_1}...x_n^{a_n}$, and $|a| = a_1 + ... + a_n$ ($|a|$ is the degree of x^a). One should notice that a constant monomial is also included and it is represented by $(0, ..., 0)$.

Definition 2.3 (monomial ordering). A monomial ordering on the set of all the monomials in $k[x_1, ..., x_n]$ is a total order on monomials that is also well founded (a well-ordering) and compatible with multiplication.

We can now define the **leading monomial**$_<$ of a polynomial p to be the greatest monomial occurring in p according to the order $<$. The most commonly used monomial orderings are lexicographical order and degree reverse lexicographical order, see [14].

Now we can define a Gröbner basis:

Definition 2.4 (Gröbner basis). Let $<$ be a monomial ordering on $k[x_1, ..., x_n]$. Let $\mathcal{I} \subset k[x_1, , x_n]$ be an ideal. A Gröbner basis of I is a finite set of generators $g_1, ..., g_m$ of $\mathcal{I}$ such that every leading monomial of a polynomial $p \in \mathcal{I}$ is a multiple of a leading monomial of a generator g_k.

Gröbner basis algorithms can be directly used to recover the solution of a set of equations (see [14]).

In the F4 algorithm, as explained above, two main stages are performed. Unlike in the simple XL algorithm, these two stages are repeated again and again (see algorithm 2.1). After Gaussian elimination is performed, some new polynomials of degree less than D may appear. We call such polynomials **degree falls**. In this case, a new expansion[D] is performed on these polynomials. On the contrary, if no such polynomial appear, the degree is incremented and an expansion[D+1] is performed. With F4, this expansion step is performed in such a way that many unnecessary polynomials are not included in the matrix before the Gaussian elimination, but we will not get into these details here.

Each expansion followed by a Gaussian elimination will be called a **step** in the F4 algorithm.

The notion of degree fall is very important in this paper. For a system $\mathcal{S}$, we call the Maximal Degree of $\mathcal{S}$ ($MD(\mathcal{S})$) the maximal D such that an Expansion[D] is necessary to recover the Gröbner basis. With both XL and F4,

---

Algorithm 2.1. F4 algorithm summary

INPUT : a system $\mathcal{S}$ of equations $\{p_1, ..., p_m\}$ describing an ideal $\mathcal{I}$
OUTPUT: a Gröbner basis of $\mathcal{I}$
$D \longleftarrow 1 + \min_{1 \leq i \leq k} degree(p_i)$
while $\mathcal{S}$ is not a Gröbner basis of $\mathcal{I}$:
    $\mathcal{E} \longleftarrow Expansion[D](\mathcal{S})$
    $\mathcal{S} \longleftarrow Gaussian\ \ elimination(\mathcal{E})$
    if $\mathcal{S}$ does not contain degree falls
        $D \longleftarrow D + 1$
return $\mathcal{S}$

---

the appearance of degree falls after a step at degree at most D is a necessary condition for $MD(\mathcal{S})$ to be D. Indeed, if the solution is unique, it has the form

$$x_1 + a_1 = 0$$
$$\vdots$$
$$x_n + a_n = 0,$$

where the $x_i s$ are the unknowns whose values are looked for and the $a_i s$ are constants: these equations are the final degree falls of degree 1. But, of course, the appearance of degree falls after a step at degree D is not a sufficient condition for the system to be solved at degree D (when $MD(\mathcal{S})$ is D).

To perform our computer simulations, we used our implementation of the algorithm F4, and another algorithm called ElimLin, see 2.2. The maximal degree of the polynomials during the computations is fixed to 2 in this algorithm. Given the very large number of variables in the original systems resulting from block ciphers, this is in fact the maximal degree for which it is conceivable to work

---

Algorithm 2.2. ElimLin algorithm

INPUT : a system $\mathcal{S}$ of GF(2) equations $\{p_1, ..., p_m\}$ of maximal degree 2
        describing an ideal $\mathcal{I}$
Apply a total order on the monomials of $\mathcal{S}$
$\mathcal{S} \longleftarrow Gaussian\ \ elimination(\mathcal{S})$
$L \longleftarrow$ Number of linear equations in $\mathcal{S}$
while $L > 0$:
    for $i = 1$ to L:
        $v \longleftarrow$ greatest variable of the linear equation l_i
        $l'_i \longleftarrow l_i \oplus v$
        Substitute v by l'_i in all the equations of $\mathcal{S}$ except from l_i
    Apply a total order on the monomials of $\mathcal{S}$
    $\mathcal{S} \longleftarrow Gaussian\ \ elimination(\mathcal{S})$
    $L' \longleftarrow$ Number of linear equations in $\mathcal{S}$
    $L \longleftarrow L - L'$
return $\mathcal{S}$

---

given that RAM and disk space in today's computers is quite limited. It appears that overall, in this paper we do not obtain better results with the F4 version of the computer algebra system MAGMA (see [13]) than with our fairly basic version of ElimLin.

3   Computer Simulations

When the number of variables or when D is large, Gaussian elimination has to be performed on huge matrices. Currently the large amount of memory needed does not allow to make simulations on many real life ciphers. This motivates the research done on toy ciphers: in these ciphers the S-boxes are quite small (3-4 bits) and the linear layer is relatively simple.

3.1   Description of Our Toy Ciphers: ToyBlock, ToyBlockS and ToyBlockR

This toy cipher is a very simple block cipher. It is inspired by Serpent, as it combines bijective 4-bit S-boxes and linear transformations. Several versions are possible (they will be called ToyBlockS and ToyBlockR). It takes blocks of $4B$ bits. The first round is the round number 1. Each ToyBlock is made of r rounds, and each round is made of:

- A bitwise XOR of the current state Z_i with the derived key, where $i + 1$ is the number of the next round. The plaintext is then Z_0. I.e:

$$X_{i+1} = Z_i \oplus K_i,$$

where K_i is the derived key and X_{i+1} the state of the inputs of the S-boxes. This gives the following for the bit j of the state, $j = 0..4B - 1$:

$$X_{i+1,j} = Z_{i,j} \oplus K_{i,j},$$

- A layer of B parallel 4 bits S-boxes. Each S-box is either the S-box number 2 of the block-cipher Serpent, see [1][1], or the reduction of the Rijndael S-box proposed by Cid et al, see [15]. The bit number j that is an output of the layer of the S-boxes of round number i is called $Y_{i,j}$.
- A linear layer. This linear layer is very similar to the one proposed in [9]. It is defined for every i in $1, ..., r$ as follows:

$$
\begin{cases}
Z_{i,1987.j+257 \mod (4.B)} = Y_{i,j} \oplus Y_{i,j+137 \mod (4.B)} \\
\qquad \text{if } j \neq 257 \mod (4.B) \\
Z_{i,1987.j+257 \mod (4.B)} = Y_{i,j} \oplus Y_{i,j+137 \mod (4.B)} \oplus Y_{i,j+274 \mod (4.B)} \\
\qquad \text{if } j = 257 \mod (4.B).
\end{cases}
$$

---

[1] Choosing a different S-box or using several different S-boxes among the Serpent S-boxes should give very similar attacks as our, yet make results more difficult to interpret.

The key schedule is a simple permutation of the key bits:

$$K_{i,j} = K_{0,(j+i \mod (4.B))},$$

where K_0 is the secret key.

We call a general version with r rounds and B S-boxes ToyBlock(r,B). If the S-box is the S-box of Serpent, we call it ToyBlockS. When it is the reduction of the Rijndael S-box of [15], it is called ToyBlockR. The Serpent S-box is a random S-box, as explained in [1], whereas the Rijndael-type one is non-random. Both can be described by a system of implicit quadratic equations.

3.2   First Simulations on ToyBlockS and ToyBlockR

We first applied F4 on ToyBlockS(3,4) and ToyBlockS(3,6). We wrote the system of equations S and S' provided by one single random plaintext6ciphertext pair for ToyBlockS(3,4) and ToyBlockS(3,6). Then we applied F4 (our implementation and the implementation of the computer algebra system Magma (see [13])) on each system. We were not able to find the solution in reasonable time, and could conclude that the necessary degrees $D = MD(S)$ and $D' = MD(S')$ are both at most 4. This degree leads to a theoretical complexity of respectively 2^{53} and 2^{73} (see section 2.2).

However, it was possible to solve the system describing 32 P-C pairs of ToyBlockS(3,4) in 3 minutes and to solve the system describing 64 P-C pairs of ToyBlockS(3,6) in 51.54 minutes with ElimLin. In this example, it is still bigger than the exhaustive search of the key. But this shows that in some cases, by using several P-C pairs, we can decrease the maximum degree of polynomials that are used in the computation from 4 to 2, and thus greatly improve the feasibility and the complexity of algebraic attacks.

We made the same computations on ToyBlockR(3,4) and ToyBlockR(3,6). We could neither recover the key by applying F4 or ElimLin on the system describing one P-C pair nor on the systems describing several P-C pairs.

Some other simulations were made on ToyBlockS(6,3) and ToyBlockR(6,3). In these simulations all the computations were made on polynomials of degree at most 2. This means that the maximal degree D in algorithm 2.1 is here 2. Our aim is not to systematically find the solution, but to look at the number of degree falls appearing. The results of these computations are given in Table 1. The third line in Table 1, called "Rounds concerned" gives the number of the rounds where some variables are implied in the new linear equations found for ToyBlockS. For example when 8 P-C pairs are used, the number of new linear equations increases and the new linear equations imply variables of the third round.

Table 1. Number of linear equations found at degree 2 for ToyBlock(6,3)

Number of P-C pairs	1	2	4	8	16	32	64
Linear eqs. for ToyBlockS(6,3)	0	15	57	151	345	740	1702
Rounds concerned ToyBlockS(6,3)	0	1, 2, 5, 6	1, 2, 5, 6	1, 2, 3, 4, 5, 6	1, 2, 3, 4, 5, 6	1, 2, 3, 4, 5, 6	1, 2, 3, 4, 5, 6
Linear eqs. for ToyBlockR(6,3)	0	0	0	0	0	0	0

The result of the comparisons between computations on ToyBlockS and computations on ToyBlockR shows that the shape of the implicit equations describing the cipher has a large influence concerning the appearance of degree falls during the computations of F4 on the systems describing this type of cipher. We analyse theses differences of shape in Section 3.3.

3.3   Analysis of the Simulations

In the next sections we will write the exclusive or operation (XOR) additively: as "+" in $GF(2)$. The ToyBlockS S-box provides 21 quadratic linearly independent equations in the input and output bits. These equations completely define the S-box. Four of them have a special property:

$$x_2 + x_3 + x_4 + x_1 x_3 + y_1 = 0$$
$$x_1 + x_2 + x_4 + x_1 x_4 + y_2 + y_3 + y_4 + 1 = 0$$

and

$$x_1 + y_1 + y_2 + y_3 + y_2 y_3 + y_2 y_4 = 0$$
$$x_1 + x_2 + x_3 + y_3 + y_4 + y_2 y_3 + y_3 y_4 + 1 = 0,$$

where the x_is are input variables and y_is output variables. The first group have degree 2 monomials only in x_i and the second group have degree 2 monomials only in y_i. This property turns out to have a huge impact on algebraic cryptanalysis. We call this kind of equations respectively "X^2 equations" and "Y^2 equations". We will show that the degree falls appearing during the simulations of section 3 are caused by this property.

Let us consider for example the toy cipher ToyBlockS(5,2). Let us consider two P-C pairs. We denote $x_{0,i}$ and $y_{0,i}$ the variables of input and output of one S-box of the first pair and $x_{1,i}$ and $y_{1,i}$ the variables of the second pair (here we only consider the leftmost S-box of the first round).

The following equations are true, the left part corresponding to the first P-C pair and the right part to the second one:

$$x_{0,1} + k_0 + z_{0,1} = 0 \qquad\qquad x_{1,1} + k_0 + z_{1,1} = 0$$
$$x_{0,2} + k_1 + z_{0,2} = 0 \qquad\qquad x_{1,2} + k_1 + z_{1,2} = 0$$
$$x_{0,3} + k_2 + z_{0,3} = 0 \qquad\qquad x_{1,3} + k_2 + z_{1,3} = 0$$
$$x_{0,4} + k_3 + z_{0,4} = 0 \qquad\qquad x_{1,4} + k_3 + z_{1,4} = 0$$
$$x_{0,2} + x_{0,3} + x_{0,4} + x_{0,1}x_{0,3} + y_{0,1} = 0 \qquad x_{1,2} + x_{1,3} + x_{1,4} + x_{1,1}x_{1,3} + y_{1,1} = 0$$
$$x_{0,1} + x_{0,2} + x_{0,4} + x_{0,1}x_{0,4} + y_{0,2} + \qquad x_{1,1} + x_{1,2} + x_{1,4} + x_{1,1}x_{1,4} + y_{1,2} +$$
$$y_{0,3} + y_{0,4} + 1 = 0 \qquad\qquad\qquad y_{1,3} + y_{1,4} + 1 = 0$$

The $z_{i,j}$ are the plaintext bits. We can choose for example $z_{0,1} = 0$, $z_{0,2} = 1$, $z_{0,3} = 1$, $z_{0,4} = 1$, and $z_{1,1} = 1$, $z_{1,2} = 0$, $z_{1,3} = 0$, $z_{1,4} = 0$. Then, by substitution we derive the following true equations:

$$y_{0,1} + k_1 + k_2 + k_3 + k_0 + k_0 k_2 + 1 = 0 \qquad y_{1,1} + k_1 + k_3 + k_0 k_2 = 0$$
$$y_{0,2} + y_{0,3} + y_{0,4} + k_1 + k_3 + k_0 k_3 + 1 = 0 \qquad y_{1,2} + y_{1,3} + y_{1,4} + k_0 + k_1 + k_0 k_3 = 0$$

The first equation of the first pair has the same degree 2 monomials than the first equation of the second pair, and this happens also for the second equations. This comes from the special structure of the equations.

If the monomial ordering is chosen such that degree 2 monomials are greater than degree 1 monomials (like degree reverse lexicographical order), after the next Gaussian elimination, we obtain linear equations between $y_{0,i}$, $y_{1,i}$ variables and some key variables:

$$y_{0,1} + y_{1,1} + k_0 + k_2 + 1 = 0$$
$$y_{0,2} + y_{0,3} + y_{0,4} + y_{1,2} + y_{1,3} + y_{1,4} + k_0 + k_3 + 1 = 0$$

Thereby we obtain linear equations mixing variables of the input of the second round of the two pairs and key variables.

The substitution used here is completely equivalent to the first step of F4, where the expansion is performed on the linear equations. Indeed, in this case, if the monomial ordering is chosen in such a way that the x_i variables are greater than the other variables, they are multiplied by other x_i variables and k_i variables during the expansion step. During the Gaussian elimination of the first step of F4, all the polynomials which leading monomials are of the type $x_i x_j$ or $x_i k_j$ are reduced with these expanded linear polynomials. In the "X^2 equations", the only remaining degree 2 monomials are of the type $k_i k_j$. Then after the Gaussian elimination we obtain the same new linear equations as those described above.

If we use 3 P-C pairs, we will obtain 6 such linear equations between the variables of the input of the second round. But only 4 of them are linearly independent. Actually, all the new equations can be parameterized by the variables of the first P-C pair and the key bits. Then by using n P-C pairs we automatically obtain $2n - 2$ new linear equations per S-box in the variables of the beginning of the second round and the key bits. For example, for ToyBlockS(5,2), we obtain $4n - 4$ such linear equations.

Now let us consider the family of the block ciphers that have the same structure as ToyBlockS, this meaning that it is made of several rounds composed of three layers: (i) XOR of the round key, (ii) layer of B parallel S-Boxes and (iii) a linear layer, followed by (i') a final XOR of the round key. We suppose to simplify that the parallel S-boxes are all identical. Then we can show the following result:

Proposition 3.4. Let us suppose the S-box is described by r quadratic equations, t of them being "X^2 equations", and t' of them being "Y^2 equations". Then after the first step of F4 applied to the equations describing n P-C pairs, at least $B(tn - t) + B(t'n - t')$ linear degree falls appear.

The proof of this proposition is given in [21].

Remark 1. Let us consider the following ratio: "number of linear equations/ number of P-C pairs" for ToyBlockS(5,2). Its value is $\frac{4n-4}{n} = 4 - \frac{4}{n}$. Then by increasing the number of P-C pairs we increase the number of new linear equations *per* P-C pair.

Those linear equations are found at the first "step" of F4 in degree 2. Of course it does not happen the same way for the other rounds at further steps of the algorithm. But by computing the other "steps", we observe experimentally that we obtain some other linearly independent degree falls. Importantly, the more P-C pairs we use, the largest is the ratio "total number of linear degree falls"/"number of P-C pairs".

We have shown that if the equations of the S-Box have some special properties, using several P-C pairs decreases the value of the maximal degree D in the algorithm F4. But as the number of variables of the system of equations increases at the same time, it is difficult to know in which cases it improves the complexity of such attacks. In Section 4, we propose some methods to exploit this weakness and show by our simulations that in some cases it improves the complexity of algebraic attacks on block ciphers of the ToyBlock type.

Remark 2 (Observation on Small Rijndael-type S-box). The equations describing the S-box of ToyBlockR does not contain any "X^2 equation" nor "Y^2 equation". This is the reason for which we are not able to improve the complexity of F4 at degree 2 on this cipher by using several P-C pairs, and the reason for which we do not obtain any degree fall either.

4   Guess-and-Determine and Chosen Plaintext Attacks

4.1   Guessing Some Key Bits

A guess-and-determine or guess-then-algebraic attack seems to be the most natural way to go further. Indeed, guessing some key bits has the same effect as the method described before: getting rapidly new linear equations. This is true for one P-C pair. For example, if four consecutive bits of secret key corresponding to the input of an S-Box are guessed, the variables of the output of the S-boxes of the first round are known. If only a part of the input key bits are guessed, some quadratic equations describing the S-Box may become linear. In both cases, new linear equations in the variables of the beginning of the second round appear.

More interestingly, guessing variables allows to gain immediately linear equations mixing variables of several different rounds. For example, if 4 key variables are guessed for ToyBlockS(4,3), we obtain 1 linear equation mixing variables of the second round and the third round with one P-C pair. With four P-C pairs, we obtain 5 such equations, and with 16 P-C pairs, we obtain 33 such equations.

Moreover, even if a part of the final complexity is purely exponential, these kind of attacks are easy to test on a PC as it suffices to fix a certain number of variables in the system of equations. The larger this number is, the easier to solve is this resulting system, in terms of time and memory complexity.

4.2 Chosen Plaintext Attacks

Method. As we use several P-C pairs, another natural direction is to try to choose carefully the plaintexts. Let us fix the value of the first plaintext. A first idea is to change only the bits corresponding to the input of one or a few S-boxes in the other plaintexts. This way, many of the other S-boxes output values of the first round are exactly the same in all the P-C pairs. Therefore we gain new very short linear equations for free.

When some key bits are guessed there is an experimentally even better method. It consists in choosing in a precise way the bits that differ between the P-C pairs at the place where the key bits xored to this part of the input are known. It means that if the key bits k_0, k_1, k_2, k_3 are guessed, the bits of plaintext corresponding to the input of the leftmost S-Box will differ, but not the other bits of plaintext. For example as we know the four key bits, it is possible to choose two plaintexts such that the output of the S-box number 1 differ only from one bit. Then, as the diffusion allows it here, only two S-boxes of the second round have a different output. We still gain equalities, thus linear equations mixing variables of the beginning of the *third* round. This method increases the number of required P-C pairs. But in our experimentations as the number of concerned S-boxes in never more than two, this number is never more than 256.

Computations on ToyBlock. We made some computations to test the efficiency of these kind of attacks. Let us explain our method. Brute force is the exhaustive search of the key. If we fix all but a key bits, an attack will be faster than brute force if the running time is less than $2^a E$, where E is the time to check one potential possibility for the key. When the number of P-C pairs is more than 2, this system has a unique solution that gives the key bits. Exact figures are hard to evaluate because they depend on an optimised implementation of the cipher. Here we will assume that one encryption takes 300 CPU clocks and that the CPU runs at 3 GHz. Then $E \approx 2^{-35}$ hours. Thus, if we fix all key bits except 35, an attack done in less than 1 hour on a PC will be faster than brute force.

By applying our substitution algorithm on the quadratic system describing 4 random P-C pairs of ToyBlockS(4,32) with 88 variables fixed, we were not able to recover the key. But by applying the same software on 4 carefully chosen P-C pairs (as described above), we could recover it in 20 minutes. We were also able to recover the key for ToyBlockS(5,32) with 16 P-C pairs and 84 variables guessed in 32 hours. Therefore it breaks ToyBlockS faster than the exhaustive search of the 128 key bits for 5 rounds with a complexity of 2^{124}.

This result shows that using several P-C pairs is a non-negligible improvement in the complexity of algebraic attacks. Indeed, our computations showed that it is completely impossible to recover the key with such a guess-then-algebraic chosen plaintext attack with the knowledge of only one P-C pair. This is an interesting development in algebraic cryptanalysis on block ciphers. In the past, the only computations on GF(2) equations describing the problem of the key recovery on toy block ciphers we are aware of, have been proposed in [15, 16], and did not allow to recover the key for systems larger than one 8-bit S-box or two 4-bit S-Boxes.

The Real-life Cipher Present. In [20], a new RFID-oriented block cipher has been presented by Bogdanov *et al.* It is a 32 rounds SPN block cipher. The block length is 64 bits and two key length of 80 and 128 bits are supported. Each of its rounds is made of a XOR of the round key K_i, a linear bitwise permutation and a layer of 16 parallel 4 bits S-boxes. All the S-boxes are identical. Compared to ToyBlockS, Present is very similar. The S-box seems to have exactly the same properties: it can be described by 21 linearly independent implicit quadratic equations, 2 of them are "X^2-equations" and 2 of them are "Y^2-equations". We note that it has a weaker linear layer as it is a simple bit permutation. Its key schedule is however stronger than ToyBlock's as a part of it is nonlinear. The authors of Present considered algebraic attacks, and explain that even considering a system consisting of seven S-boxes, i.e. a block size of 28 bits, they were unable to get a solution in a reasonable time to a two-round version of the reduced cipher [20]. We implemented the system of equations corresponding to 64 P-C pairs of the real cipher reduced to 5 rounds and by fixing the first 12 bits of the key and 20 bits of roundkey (the 4 last bits of each round key except the first one) and we were able to recover the solution in 1.82 hours. This breaks 5-round reduced version of the 80 key-bits Present cipher.

It is far from breaking the real cipher, but shows that the authors have not thought about algebraic attacks that use of several P-C pairs concerning this type of attacks. It appears also that they did not consider guess-then-algebraic attacks.

5   Towards New Design Criteria for S-Boxes?

First we prove that "X^2 equations" and "Y^2 equations" always exist for 3 bits S-boxes. This comes from the following theorem:

Theorem 5.1 ([5]). For any $n \times m$ S-box, $F : (x_1, \ldots, x_n) \mapsto (y_1, \ldots, y_m)$, and for any subset T of t out of 2^{m+n} possible monomials in the x_i and y_j, if $t > 2^n$, there are at least $t - 2^n$ linearly independent I/O equations (algebraic relations) involving (only) monomials in T, and that hold with probability 1, i.e. for every (x, y) such that $y = F(x)$.

Application: For $n = m = 3$, the dimension of the space T of all equations of type

$$\sum \alpha_{ij} x_i x_j + \sum \beta_i x_i + \sum \gamma_i y_i + \sum \delta = 0$$

is $1 + 3 + 3 + 3 = 10$ while $2^n = 8$. Thus there are always at least $2 = 10 - 8$ of "X^2 equations". By the same argument, there are also always at least two "Y^2 equations".   □

Remark 3. This result explains why the fast algebraic attacks on CTC from [8] are so efficient. Indeed, the CTC S-box can be represented by 14 linearly independent implicit quadratic equations, 3 of them being "X^2 equations" and 3 of them being "Y^2 equations". By using several P-C pairs, many linear degree falls are obtained at the beginning of the computation.

5.1   Larger S-Boxes and Resulting Security Recommendations

We cannot prove the same result for 4 bits S-boxes. Many random S-boxes like all of the 8 Serpent S-boxes have "X^2 equations" and "Y^2 equations". However, certain S-boxes have no such equations, for example the Rijndael-type S-box on 4 bits. Still, this only protects against one algebraic attack at degree 2. What about higher degrees? Results similar to the above theorem, can be quite easily shown for higher degree equations. For example at degree 3, if we have "X^3 equations", this meaning that the cubic monomials of the equation only mix variables of the input of the S-box, then we have degree falls (of degree 2) implying variables of several P-C pairs. But it is very difficult to measure the impact of such equations on algebraic attacks with several P-C pairs, as in this case the obtained degree falls have degree 2 instead of 1. For example, we computed and obtained as many as 47 "X^3 equations" and 48 "Y^3 equations" for the AES S-box on 8-bits. Lack of theory, and insufficient computing power available, make it difficult to evaluate the importance of these equations in attacks on reduced-round AES.

S-box Criteria. Thus it seems that a desirable property for an S-box is just to avoid "X^2 equations" and "Y^2 equations". Indeed, it allows to avoid linear relations between variables of several P-C pairs. Our study shows that the Serpent and Present S-boxes are more vulnerable to algebraic attacks than the AES S-box which does not have any "X^2" nor "Y^2 equations". But Serpent and Present have both 32 rounds, the AES has only 10 (which is already a lot w.r.t. to our attacks anyway). In general, algebraic attacks are still too badly theoretically understood to make meaningful extrapolations for a larger number of rounds.

6   Conclusion

This paper initiates the study of specific S-box criteria that make algebraic key recovery attacks much more efficient when several plaintext-ciphertext pairs are used. These S-box criteria are based on a particular type of the S-box multivariate equations, that we showed to produce many linear degree falls during the computation of Gröbner bases algorithms suc as XL, F4, F5 or the simplest possible - ElimLin. We showed that all 3-bit S-boxes are vulnerable, i.e. whatever is the S-box used. For S-boxes on 4 bits, the weakness is no longer systematic.

On the one hand, toy versions of Serpent and the lightweight cipher Present are shown to be vulnerable. On the contrary to what the designers of Present have claimed, we showed that it is possible to break 5 rounds of Present with a simple guess-then-algebraic attack with a very low quantity of plaintext/ciphertext pairs. A similar attack was obtained for ToyBlockS. In real life, these ciphers have as many as 32 rounds. This is quite far away from what experimental algebraic cryptanalysis can ambition to break in a near future.

On the other hand, it appears that the 4-bits version of a Rijndael-type S-box has no equations of the types we exploit in this paper. We showed that it is then much more resistant than many other S-boxes of the same size against this type of algebraic attacks. The real AES S-box seems to be also very resistant against these attacks.

References

1. Anderson, R., Biham, E., Knudsen, L.: Serpent, a flexible Block Cipher With Maximum Assurance First AES Candidate Conference, Ventura California (1998), http://www.cl.cam.ac.uk/~rja14/serpent.html
2. Ars, G., Faugère, J.-C., Sugita, M., Kawazoe, M., Imai, H.: Comparison between XL and Gröbner Basis Algorithms. In: Lee, P.J. (ed.) ASIACRYPT 2004. LNCS, vol. 3329, pp. 338–353. Springer, Heidelberg (2004)
3. Courtois, N.: The Inverse S-box, Non-linear Polynomial Relations and Cryptanalysis of Block Ciphers. In: Dobbertin, H., Rijmen, V., Sowa, A. (eds.) AES 2005. LNCS, vol. 3373, pp. 170–188. Springer, Heidelberg (2005)
4. Courtois, N., Bard, G.V., Wagner, D.: Algebraic and Slide Attacks on KeeLoq, preprint (2007), http://eprint.iacr.org/062/
5. Courtois, N. Bard G.V.: Algebraic Cryptanalysis of the Data Encryption Standard. In: Cryptography and Coding, 11-th IMA Conference, Cirencester, UK (December 18-20, 2007) (to appear), eprint.iacr.org/2006/402/; Also presented at ECRYPT workshop Tools for Cryptanalysis, Krakow (September 24-25, 2007)
6. Courtois, N., Pieprzyk, J.: Cryptanalysis of Block Ciphers with Overdefined Systems of Equations. In: Zheng, Y. (ed.) ASIACRYPT 2002. LNCS, vol. 2501. pp. 267–287. Springer, Heidelberg (2002)
7. Courtois, N., Shamir, A., Patarin, J., Klimov, A.: Efficient Algorithms for solving Overdefined Systems of Multivariate Polynomial Equations. In: Preneel, B. (ed.) EUROCRYPT 2000. LNCS, vol. 1807. pp. 392–407. Springer, Heidelberg (2000)
8. Courtois, N.T. : How Fast can be Algebraic Attacks on Block Ciphers? In: Biham, E., Handschuh, H. Lucks, S. Rijmen, V. (eds.) Proceedings of Dagstuhl Seminar 07021, Symmetric Cryptography (January 2007), http://drops.dagstuhl.de/portals/index.php?semnr=07021, http://eprint.iacr.org/2006/168/ ISSN 1862 - 4405
9. Courtois, N.: CTC2 and Fast Algebraic Attacks on Block Ciphers Revisited, http://eprint.iacr.org/2007/152/
10. Dunkelman, O., Keller, N.: Linear Cryptanalysis of CTC, http://eprint.iacr.org/2006/250/
11. Faugère, J.-C.: A new efficient algorithm for computing Gröbner bases (F_4). Journal of Pure and Applied Algebra 139, 61–88 (1999), www.elsevier.com/locate/jpaa
12. Faugère, J.-C.: A new efficient algorithm for computing Gröbner bases without reduction to zero (F5). In: Workshop on Applications of Commutative Algebra, Catania, Italy, April, 3-6. ACM Press, New York (2002)
13. MAGMA, High performance software for Algebra, Number Theory, and Geometry, — a large commercial software package, http://magma.maths.usyd.edu.au/
14. Buchberger, B., Winkler, F.: Gröbner Bases and Application. London Mathematical Society, vol. 251. Cambridge University Press, Cambridge
15. Cid, C., Murphy, S., Robshaw, M.J.B.: Small Scale Variants of the AES. In: Gilbert, H., Handschuh, H. (eds.) FSE 2005. LNCS, vol. 3557. pp. 145–162. Springer, Heidelberg (2005)
16. PhD Thesis, http://gwenole.ars.ifrance.com/
17. Cid, C., Leurent, G.: An Analysis of the XSL Algorithm. In: Roy, B. (ed.), ASIACRYPT 2005. LNCS, vol. 3788, pp. 333–352. Springer, Heidelberg (2005)

18. Buchmann, J., Pyshkin, A., Weinmann, R.-P.: Block Ciphers Sensitive to Gröbner Basis Attacks. In: Pointcheval, D. (ed.) CT-RSA 2006. LNCS, vol. 3860. pp. 313–331. Springer, Heidelberg (2006)
19. Lim, C.-W., Khoo, K.: Detailed Analysis on XSL Applied to BES. In: Biryukov, A. (ed.) FSE 2007. LNCS, vol. 4593. pp. 242–253. Springer, Heidelberg (2007)
20. Bogdanov, A., Knudsen, L.R., Leander, G., Paar, C., Poshmann, A., Robshaw, M.J.B., Seurin, Y., Vikkelsoe, C.: PRESENT: An Ultra-Lightweight Block Cipher. In: Paillier, P., Verbauwhede, I. (eds.) CHES 2007. LNCS, vol. 4727. pp. 450–466. Springer, Heidelberg (2007)
21. Debraize, B.: Versailles University, France, PhD Thesis, (to be published, 2008)

Combiner Driven Management Models and Their Applications

Michael Beiter

Universität Tübingen
Wilhelm-Schickard Institut für Informatik
Sand 14, 72076 Tübingen, Germany
beiter@informatik.uni-tuebingen.de
http://www.michael.beiter.org

Abstract. In this paper, we will study the question of performing arbitrary updates in secret sharing schemes when shares of unaffected parties shall remain unchanged. We will introduce a new phase in the lifetime of secret sharing schemes to simplify constructions of multi-time secret sharing schemes and propose management models that allow unlimited updates from arbitrary schemes without need for broadcasts. As an example, we give an implementation based on Shamir's threshold scheme.

Keywords: secret sharing, management models, multi-time secret sharing, threshold schemes with update capabilities.

1 Introduction

To protect a secret from abuse or to add redundancy for increased reliability, it is common practice to use secret sharing schemes, in particular threshold schemes as introduced by Shamir [11] and Blakley [3].

Setting up a secret sharing scheme is a rather expensive process, as secure channels for share distribution are obligatory: enrollment of new participants who were not issued with any private information at system setup is impossible in a broadcast only network. The reason for this is that each new participant needs to receive some private information from one of the scheme's parties. This may be the dealer, but it can also be one of the existing shareholders in so called *dealer free environments* (see *e.g.* [10] and others).

To minimize these setup costs, most secret sharing schemes are designed for quite a long lifetime. During a scheme's lifetime, it is to be expected that demand for enrollment or disenrollment of participants may arise. In some cases, it even may be required to alter the parameters of a threshold scheme or the access structure of a general secret sharing scheme.

The trivial solution for enrollment as well as disenrollment or updating is to design a new scheme and redistribute new shares. Usually, however, this is not an option. Especially in urgent cases, *e.g.* when a share is known to be disclosed or an immediate change in the access structure shall be enforced, there must

S. Lucks, A.-R. Sadeghi, and C. Wolf (Eds.): WEWoRC 2007, LNCS 4945, pp. 114–126, 2008.
© Springer-Verlag Berlin Heidelberg 2008

exist a way to act fast. Hence, the question of how to alter the access structure of secret sharing schemes is of importance.

In this paper, we will study the question of performing arbitrary updates in secret sharing schemes. We will introduce a new phase in the lifetime of secret sharing schemes to simplify constructions of multi-time secret sharing schemes. Moreover, we will propose new management models where two trusted entities are present: the dealer, who knows the shared secret, and the combiner, who supports the recreation process. In the proposed models, the dealer remains active during runtime and maintains a secure connection to the combiner. These models allow unlimited updates of arbitrary secret sharing schemes without need for broadcasts. Most importantly, the shares of unaffected parties remain unchanged in update operations. As an example, we give an efficient implementation based on Shamir's threshold scheme which is ideal from a shareholder's point of view.

2   Related Work

2.1   Dynamic Secret Sharing

Threshold schemes with disenrollement capabilities were introduced by Blakley *et al.* [4], and have been discussed in several publications (*e.g.* [5] [6] [8]). A threshold scheme protects a secret g among v participants so that i participants, $u \leq i \leq v$, can recreate g and less than u participants can't uniquely determine g. Barwick *et al.* deal with the question of altering the parameters of threshold access structures in [1] and [2]. They call changes of u and v in a (u, v) threshold scheme an *update* of the scheme. Updates can be activated by broadcast messages. However, these schemes are dedicated solutions that enlarge the participants' shares.

We call secret sharing schemes that allow unlimited updates of their access structure *fully dynamic secret sharing schemes*.

The updating of shares is related to, but at the same time different, from the updating of access structures: as the latter changes the set of participants, updating of shares keeps the access structure unchanged. Updating a certain share is useful when, for instance, a shareholder lost his share but still should be able to join recreations. This operation is possible when the scheme provides the ability to disenroll and add participants to active schemes: the disenrollment operation disables the lost share and adding a new participant allows to assign the affected participant a new one.

However, there may exist schemes which do not provide the disenrollment or enrollment operations or which can carry out share updating more efficiently. As a consequence, we separate the issue of updating of shares from enrollment and disenrollment.

2.2   Components of a Secret Sharing Scheme

Usually, secret sharing schemes are based on three main components:

1. The set of participants and the access structure
   Defining the access structure based on the power set of the participants set
   is common in research (*e.g.* [15]):

   Definition 1. *Let T be a set of participants with n_T shareholders $t_1, \ldots, t_{n_T}$
   and $\mathcal{P}(T)$ the power set of T. Furthermore let K be the set of all shares taken
   from $\mathcal{K}$ which represent the set of all possible shares.*

   *For each secret $g \in \mathcal{G}$, where $\mathcal{G}$ is the set of all possible secrets, the set
   of participant sets $\mathcal{Z} \subseteq \mathcal{P}(T)$ which can recreate g is called* access structure
   of g. The elements $Z_i \in \mathcal{Z}$ are called authorized subsets *concerning $\mathcal{Z}$. The
   remaining subsets $\bar{\mathcal{Z}} := \mathcal{P}(T) \setminus \mathcal{Z}$ are called* unauthorized subsets *concerning
   $\mathcal{Z}$. An access structure $\mathcal{Z}$ contains $n_{\mathcal{Z}}$ authorized subsets $Z_1, \ldots, Z_{n_{\mathcal{Z}}}$.*

   Usually $\mathcal{Z}$ is assumed to be *monotone*, that is $Z \in \mathcal{Z}$ and $Z \subseteq Z' \subseteq T$
   implies $Z' \in \mathcal{Z}$.

2. The distribution function with a collection of supported operations
   The distribution function according to Stinson ([15]) assigns shares to the
   involved parties so that recreating the secret using a given set of shares is
   possible if and only if the set of shares conforms with an authorized subset.
   Informally, a distribution function and a scheme based on this function is
   perfect if the probability of an unauthorized subset to guess a protected secret
   is not increased through the knowledge of their share set ([15]). Distribution
   functions which are based on certain assumptions like the difficulty of the
   discrete logarithm problem are called *conditionally secure*, and so are the
   schemes based on these functions.

   Each distribution function supports a certain set of operations. This
   includes five obligatory base operations: to initialize the secret g and a share
   k_i, to map and transfer a share k_i to a participant t_j and to recreate g
   using a set of provided shares. Depending on the distribution function, there
   may be additional operations available. These include, but is not limited to
   operations for updating both a shared secret and a participant's share or to
   update the parameters of a threshold scheme.

3. The management model with a set of supported operations
   A management model describes the infrastructure that is available in the
   secret sharing scheme. This includes for instance the type of parties involved,
   and the nature of their connections. The model usually distinguishes various
   phases, in which the parties may interact with each other using the available
   links.

   In [9], Martin gives an overview of a number of different models which
   allow dynamic access structures for unconditionally secure secret sharing
   schemes. He distinguishes two phases: in the *initialization* or *setup phase*, the
   dealer computes the shares and distributes them among the participants. In
   the *running phase*, actions like access structure updates or secret recreation
   can be performed.

   Martin lists twelve management models and categorizes them with respect
   to presence/non-presence of the dealer during the setup and running phase

and the type of connection between dealer and participants in these phases. Not all of these models are capable of accomplishing every type of access structure change.

Each model must implement a minimum set of parties and connections. This includes the participants and, at least in perfect secret sharing schemes, some secure connection in the initialization phase.

2.3   The Combiner as an Entity in Secret Sharing

In the case of what we call a public recreation, all participants attending the recreation process pool their shares and compute the secret. This form of recreation actually is what Shamir proposed [11], but broadcasting their shares and trusting other participants to honestly do the same is far too credulous: this approach is higly vulnerable against cheating (*e.g.* [14], [17])).

Schemes that keep the shares safe (*e.g.* computationally secure schemes like [6], [7] and others) only disclose the protected secret during recreation. However, in public recreation, both shares and secret are revealed.

In any case, to construct a *multi-time secret sharing scheme* that shall be used several times, it is necessary that both the participants' shares and the recreated secret remain private at all times. While the former can be achieved with various techniques, the latter is only possible with a trusted authority which performs the recreation and triggers a predefined action.

It is common practice in secret sharing to use a so called combiner as an entity which recreates the secret and triggers a certain action. Although there are schemes that do not rely on a combiner (*e.g.* [7]), this entity usually is implicitly present. However, in most established secret sharing schemes, the term "combiner" is used rather abstractly: in these schemes, the combiner can be any arbitrary entity like a not further specified third party, a tamper proof device, a group of participants in case of public recreation or even a single one of the shareholders. However, there are usually no further statements as to who the combiner is and what his abilities are.

Simmons proposed *prepositioned secret sharing schemes*, in which the dealer computes an activation secret during initialization ([12],[13]). This activation secret does not need to be published, but can be joined with a tamper proof device: the scheme becomes available when the tamper resistant module is activated.

This allows creation of schemes with some sort of veto capabilities: the combiner receives an additional share that is obligatory for recreating the secret but is not sufficient to recreate the secret. Even more, using the combiner's secret information and some shares provided by the participants, the secret can be recreated if and only if the provided shares represent an authorized subset of the access structure.

One can think of various applications based on this type of infrastructue:

- Schemes can be protected against *wildcat recreations* where an authorized subset of corrupt participants meets and recreates the secret g. As the collection of distribution functions is known, they can compute shares of other participants, which opens the way to attacks like identity theft.

- A related approach is the creation of a true "single try secret sharing scheme" to prevent cheating: in a recreation attempt, a cheater may learn about the shares of other shareholders, prevent the recreation and recreate the secret for himself if an authorized subset was present. Instead of using the known counteractive measures against cheating, the combiner can enforce a single recreation attempt by destroying his mandatory share after the first recreation attempt.

It is worth noting that the dealer, say the authority that is totally aware of the protected secret and any other information necessary for constructing and altering the scheme, can never be the combiner, as this would run contrary to the point of secret sharing: in a secret sharing scheme, we have no information sufficient for itself stored at any publicly accessible point during the running phase. It is fundamental that the private information in form of the secret is computed at the time of recreation.

The combiner in turn differs fundamentally from the dealer: to add new shareholders, the dealer must know which shares he has already assigned to and which secrets he needs to share. The combiner in contrast must prevent an accumulation of knowledge: when a secret sharing scheme is run as a multi-time secret sharing scheme, accumulation of private shares and recreated secrets is unavoidable during recreation. The combiner as a trustworthy entity is characterized by the fact that he destroys secret knowledge as soon as possible.

As the combiner handles the recreation, triggers a predefinied action and usually keeps the participants' shares k_i and the recreated secret g private, the shareholders have to trust him only insofar as he does not reveal their shares (as would be the case in conventional public recreation) and actually triggers the action as expected, which is a moderate level of trust.

There are models in which the dealer remains active after the initialization phase, e.g. to alter the access structure during runtime ([9]), which is an exception to the paradigm of an non existent single point of failure. However, the dealer assures that this private information is protected at its best, which includes that the place of storage is not publicly accessible. This is contrary to the requirements of an efficient recreation, where public access is obligatory. Hence, in schemes relying on these models, the combiner and the dealer also need to be separated.

In known schemes, the combiner e.g. as a tamper proof device and the dealer are separated in time: the dealer distributes shares to the participants and the combiner, but is no longer available during runtime. In the following, we will discuss models where both the dealer and the combiner can always be present.

3   Combiner Driven Management Models

3.1   A Third Phase in the Life Cycle of Secret Sharing Schemes

We now refine Martin's categorization of phases in the life cycle of secret sharing schemes ([9]) insofar as we propose a *recreation phase* with secure or broadcast

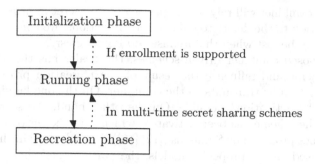

Fig. 1. Phase change possibilites

connections between the parties. As we will see in the following, multi-time secret sharing schemes in particular are only possible when the recreation phase with secure connections is explicitly present. For conventional single-time secret sharing schemes, the always implicitly present recreation phase with broadcast connections between the participants is sufficient.

In multi-time secret sharing, it is necessary to switch between the running and the recreation phase: in a recreation attempt, the phases change from running over recreation back to running. Thus we can define the life cycle of a secret sharing scheme as follows (see figure 1):

- A newly created scheme starts in the initialization phase: the secret is determined, the distribution function is selected and the participants are initialized with their shares. Afterwards, the scheme switches to the running phase.
- Depending on the model's and the distribution function's abilities, the access structure can be updated during the running phase. We define that the scheme switches into the initialization phase during enrollments, as the enrollment of shareholders is an initialization process.
- As soon as a (not necessarily authorized) subset of participants requests a recreation, the scheme switches into the recreation phase. Depending on the security policy and on whether the scheme is a single- or multi-time secret sharing scheme, the scheme's lifetime may end at this point, no matter whether the recreation was successful or not:

    The lifetime of single-time secret sharing scheme always ends after the first successful recreation at the latest. In multi-time secret sharing schemes, the scheme may switch back into the running phase after a recreation attempt.

3.2 The Combiner as a Trusted Party

We propose new management models that allow creation of schemes which can perform general updates as described in chapter 2.1. In the proposed models, the combiner is an additional trusted entity which has a secure link to the dealer and can optionally store a certain amount of private information. The dealer, who knows the shared secret, is active in both the initialization and the running phase. As the combiner will rely on the privacy of his secret information and the secure connection to the dealer to enforce certain actions, we concede that these capabilities may be lost when the premises no longer exist.

In the proposed combiner driven schemes, the dealer has the ability to activate the scheme and influence the result of the recreation process with the private information he transmits to the combiner. On the one hand, this makes the scheme's "deactivation" possible. On the other hand, the secret g can be altered after the scheme has been activated, which is an extension of the prepositioned schemes proposed by Simmons ([12],[13]). We will show in chapter 4 that in schemes based on the proposed models, the participants' shares can remain unchanged in update operations.

There are two variants of the proposed model which are both based on three parties: the dealer, the combiner and the participants. Each of these three parties is present in the intialization, running and recreation phase. However, their level of activity and the type of connections between the parties vary:

- During the initialization phase, the dealer has secure connections to the participants and the combiner. The combiner and the participants cannot initiate any communcation in this phase.
- In the running phase there is only a secure link between the dealer and the combiner. No other communication of any kind can take place.
- The dealer is not active during the recreation phase and hence he has no connection to the combiner. The sort of link between the combiner and the participants depends on the variant of the model:
  - In variant A, the combiner has a secure connection to the participants.
  - In variant B, the combiner can only communicate with the participants via broadcast messages.

Note that there is never any direct communication between the participants in model A and B.

Variant A is the more powerful model as the combiner has a secure connection to the participants during the recreation phase. This model allows the creation of multi-time secret sharing schemes, as the shares as well as the recreated secret remain private at all times. The combiner can drop the multi-time property by downgrading the secure connection during recreation from a secure to a broadcast connection. This results in a model switch to model B. If the combiner does not store any private information, but is just a trusted entity used for recreation and so gaining a multi-time secret sharing scheme, the participants can decide to recreate the secret by public recreation. This is reasonable when the combiner (and some cloned instances if applicable) is unavailable $e.g.$ due to a denial of service attack. The decision to carry out a public recreation results in a model change from model A to a conventional model where neither a dealer nor a combiner are present during recreation and only broadcast connections between the participants are available.

In variant B, the combiner only has a broadcast connection to the shareholders. Schemes based on this model are limited to one time secret sharing: we assume that a connection cannot be "upgraded", $i.e.$ a broadcast connection shall not become a secure connection, as this would not make sense in reality. To protect schemes settled in this model against wildcat recreations, the combiner here usually holds some private information which is vital for recreating the secret.

To summarize, the secure connection between combiner and participants during the recreation phase is optional and only necessary to provide multi-time secret sharing. The secure connection between dealer and combiner allows updating of private information on the part of the combiner during runtime. This can be used to create fully dynamic secret sharing schemes. The rather strong requirement of a secure connection between dealer and combiner ist softened in practice, as it is needed for updating only. The connection's type may vary from temporary physical contact up to a full featured public key infrastructure.

4   A Fully Dynamic Secret Sharing Schemes Realizing Threshold Access Structures

In the following, we will give an implementation for a secret sharing scheme that realizes threshold access structures in combiner driven management models. Let $T = \{t_1, \ldots, t_v\}$ be the set of participants and $\mathcal{P}(T)$ the power set of T. Then $\mathcal{Z} = \{A \subseteq \mathcal{P}(T) : |A| \geq u\}$ is the access structure of a (u, v) threshold scheme $(u \leq v)$.

The proposed construction is based on Shamir's threshold scheme as described in [11], with the difference that a polynomial of a larger degree will be used. This modification of Shamir's scheme allows general updates of the access structure, which includes enrollment, disenrollment and changes of the threshold. Furthermore, the shared secret can be updated in active schemes.

It is worth mentioning that Shamir's original scheme can benefit *e.g.* from multi-time secret sharing or wildcat recreation protection by settling it in a combiner driven model.

Construction. Let $\mathbb{F}_p$ be a finite field of prime order p, $\mathcal{G} = \{0\} \times \mathbb{F}_p$ the set of all possible secrets and $\mathcal{K} = \mathbb{F}_p \times \mathbb{F}_p$ the set of all possible shares. Further, let $g = (0, y_g) \in \mathcal{G}$ be the secret and $k_i \in K \subset \mathcal{K}$ the share of party t_i, $1 \leq i \leq v$. For a (u, v) threshold access structure, we create a polynomial f of degree v instead of degree $u - 1$ as Shamir does in [11].

The dealer determines the shares $k_i = (x_i, y_i)$, $1 \leq i \leq v$ by choosing the y_i randomly and the $x_i \neq 0$ pairwise distinct. Afterwards, he interpolates the v points and the secret point $(0, y_g)$ by a polynomial f using the Newton method. In the unlikely case that $deg(f) \neq v$, the dealer restarts the procedure.

When $deg(f) = v$, the dealer computes $v - u + 1$ pairs $(c_i, f(c_i))$, $1 \leq i \leq v - u + 1$, $c_i \neq 0, x_1, \ldots, x_v$, and passes them to the combiner. The shares $k_1, \ldots, k_v$ are distributed to the participants via a protected channel, the dealer securely stores the Newton polynomial f.

For numerical efficiency, the dealer uses Newton interpolation instead of Lagrange interpolation, for instance, to calculate g and $k_1, \ldots, k_v$. Using Newton interpolation requires the dealer to evaluate $O(v^2)$ divisions to interpolate the polynomial. However, in contrast to Lagrange interpolation, the dealer computes so called *divided differences* as intermediate results. In later steps, the dealer can reuse these divided differences to evaluate the polynomial at a certain point in $O(v)$ steps. Most importantly, adding a new participant, which amounts to interpolate a polynomial through an additional point, can be achieved with only $O(v)$ operations. Hence, this approach is also more efficient in recreation or updating operations.

However, the dealer must store the divided differences, which increases the amount of required storage space by $O(v^2)$. A detailed discussion of Newton interpolation can be found for instance in [16] pp. 43 ff.

Recreation. The recreation of the shared secret is only possible with assistance of the combiner: no authorized subset of u participants or more has enough shares to interpolate f and hence to determine g. Depending on the number of parties initiating the recreation process, the combiner will contribute at least one and up to $v - u + 1$ shares, interpolate the polynomial f and compute the secret information $f(0) = y_g$. This has the advantage that the combiner can check that no wildcat recreations occur, which assures that g is kept secret under any circumstances.

Since the combiner contributes up to $v - u + 1$ shares, some steps can be precomputed to calculate the Newton polynomial: having received the additional shares $(c_i, f(c_i))$ from the dealer, the combiner precomputes the first $v - u + 1$ steps of the Newton polynomial and caches this as an intermediate result. Using the precomputed part of the Newton polynomial, the combiner can compute the polynomial using u shares and check any remaining parties by insertion.

General Updates of Threshold Access Structures. To answer the question of how to carry out general updates on threshold access structures, we will have a look at the six possible operations. We will show now that in combiner driven schemes not only enrollment or disenrollment operations but also updates of shares and secret as well as changes in the security policy are possible:

1. Participants decrease from v to $\tilde{v}$ and $\tilde{v} < v$
2. Participants increase from v to $\hat{v}$ and $\hat{v} > v$
3. Threshold decrease from u to $\tilde{u}$ and $\tilde{u} < u$
4. Threshold increase from u to $\hat{u}$ and $\hat{u} > u$
5. Updating shares in active schemes
6. Updating the secret in active schemes

We presume that the combiner as a trusted entity follows the protocol correctly. This includes in particular that the combiner honestly follows the dealer's instructions to delete or replace, respectively, the shares under his control.

1. Participants' Decrease. To disenroll a participant t_k from all authorized subsets, it must be assured that his share (x_k, y_k) is no longer a point of the graph of f. Therefore, we create a new polynomial f^* that has all points (x_i, y_i), $0 \leq i \leq v$, in common with f except the point (x_k, y_k). To achieve this, the dealer chooses y_k^* randomly in $\mathbb{F}_p$ and interpolates the Newton polynomial f^* with the new (x_k, y_k^*) instead of (x_k, y_k).

Since the disenrolled user is not removed from the scheme but only deactivated, the degree of the polynomial remains v. So the dealer checks for $deg(f^*) = v$, computes $v - u + 1$ pairs $(c_i, f^*(c_i))$, $1 \le i \le v - u + 1$, which replace the corresponding former pairs $(c_i, f(c_i))$ and passes them to the combiner. The dealer does not have to inform the parties of the disenrollment, nor is there any need to communicate with the removed participant: the shares k_i of the remaining participants are unchanged, and the share $k_k = (x_k, y_k)$ becomes invalid after the combiner has updated his data.

The dealer can disenroll multiple parties by chosing multiple new polynomial values for the corresponding shares. It is easy to see that using this combiner driven management model, there are no limitations regarding any maximum disenrollment capacity or even any changes in u or v.

In the lifetime of a secret sharing scheme, the dealer should touch the share of each party only for creation and revocation: if a participant is disenrolled several times, there is indeed a small chance that the randomly chosen pair (x_k, y_k^*) matches the original (x_k, y_k) and reactivates the share. In order to circumvent this, the dealer must keep records of deactivated parties.

2. Participants' Increase. Adding one or more new parties is straight forward: as v increases to $\hat{v} > v$, the dealer needs to recompute the polynomial f, as its degree must also increase by $\hat{v} - v$.

The dealer chooses $\hat{v} - v$ pairs $(x_i, y_i) \in \mathbb{F}_p \times \mathbb{F}_p$, $v + 1 \le i \le \hat{v}$, where y_i is random, the x_i are pairwise distinct and $x_i \ne 0$, $x_1, \dots x_v$. Using the stored Newton polynomial, the dealer computes a new Newton polynomial f^* using the additional shares. This does not affect any of the existing shares $k_1, \dots, k_v$ or the secret g. As stated above, the dealer must choose different values for the new y_i if $deg(f^*) \ne \hat{v}$.

As in the initial construction, the dealer computes $\hat{v} - u + 1$ pairs $(c_i, f^*(c_i))$, $1 \le i \le \hat{v} - u + 1$, $c_i \ne 0$, $x_1, \dots x_{\hat{v}}$, and passes them to the combiner. The $\hat{v} - v$ created shares $k_{v+1}, \dots, k_{\hat{v}}$ are distributed to the new participants via a secure channel, the Newton polynomial f^* is stored by the dealer.

Trying to keep the degree of the polynomial small by reusing x_i from disenrolled shareholders is possible for non-perfect schemes: the dealer could assign an "unused" x_i from a disenrolled shareholder t_i together with a new, random y_j to a new participant t_j. However, each time a x_i is reused, the probability of assigning a y_j which has already been used for a past participant holding x_i increases. Another security concern for reusing is that a group of m former holders of a particular x_i could conspirate and so decrease the set of possible y_j by m to $p - m$. Indeed, this is no issue in practice, as p will usually be much larger than m.

If no secure channels are present, the enrollment operation is not available.

3. Threshold Decrease. The success of a recreation attempt depends on the number of shares the combiner can contribute: as described before, the combiner contributes $1 \leq s \leq v-u+1$ shares, if $v-s+1$ participants attend the recreation process.

To decrease the threshold u to a $\tilde{u} < u$, the number of shares the combiner can contribute needs to be increased to $v-\tilde{u}+1 > v-u+1$: the dealer computes $v - \tilde{u} + 1$ pairs $(c_i, f(c_i))$, $1 \leq i \leq v - \tilde{u} + 1$, $c_i \neq 0, x_1, \ldots x_v$, and transmits them to the combiner via a secure channel.

As the combiner already holds u shares, this operation can be performed with a relatively low bandwidth consumption of $u - \tilde{u}$ shares: it is sufficient to merely transfer only a set of additional shares. This implies that the dealer keeps records on which shares he has already made avaliable to the combiner.

4. Threshold Increase. According to the operations required for threshold decrease, an increase of u to $\hat{u} > u$ can be performed by decreasing the number of shares hold by the combiner to $v-\hat{u}+1 < v-u+1$. To achieve this, the dealer can either compute $v - \hat{u} + 1$ pairs $(c_i, f(c_i))$, $1 \leq i \leq v - \hat{u} + 1$, $c_i \neq 0, x_1, \ldots x_v$ and transmit them to the combiner via a secure channel or issue a delete command making the combiner invalidate $\hat{u} - u$ shares.

While the former is a rather bandwidth consuming operation, the latter can be performed at comparatively low cost: the dealer needs to keep track of the shares he assignes to the combiner. To invalidate a number of shares, it is sufficient for the dealer to transmit a delete request for the required number of shares. To keep the dealers' share records valid, the combiner deletes for instance the first $\hat{u} - u$ shares. This allows a small message footprint of $\log p$ bits. In comparison, issuing a set of new shares for the combiner requires the transmission of a message of size $(v - \hat{u} + 1) \cdot 2 \log p$ bits.

5. Updating shares in Active Schemes. In the described scheme updating of shares can be conducted by slightly modifying the disenrollment operation: after the disenrollment is complete, the new randomly chosen share is not destroyed but transmitted via a secure channel to the shareholder.

6. Updating the Secret in Active Schemes. In this scheme, the operation of updating the secret is closely related to the question of disenrolling shareholders in active schemes. To update the secret, we slightly alter the disenrollment process: the dealer choses a new secret $g^* \neq g$ in $\{0\} \times \mathbb{F}_p$ whereas the participants shares remain unchanged. Then he interpolates a new polynomial f^* of degree v with the new $(0, g^*)$ instead of $(0, g)$. Finally, the dealer calculates a new share set for the combiner by insertion. After transfering these $v - u + 1$ shares via a secure connection to the combiner, the new secret is active.

Final Considerations. It is easy to see that the proposed scheme is perfect if, as mentioned in paragraph 2., no x_i of disenrolled participants are reused. To proof this, the combiner is considered an extra participant with multiple shares.

An evaluation of the scheme's efficiency according to Stinson's definition ([15]) results in a shareholder information rate of 1. However, the combiner's information rate will usually be smaller. That is why the proposed scheme is generally not ideal.

5 Conclusions

In this work, we proposed a recreation phase as a new phase in the lifetime of secret sharing schemes to simplify the construction of truly multi-time secret sharing schemes. We introduced the combiner as a trustworthy entity with a secure connection to the dealer during runtime and the ability to store a certain amount of private information. The secure connection to the dealer, who knows the shared secret, allows strong update operations like changes in the security policy of threshold schemes while the participants' share size can remain ideal. Most importantly, the shares of unaffected parties remain unchanged in update operations. The rather strong requirement of a secure connection between dealer and combiner is softened by the fact that this connection is only obligatory during update operations. Update operations can be enforced without the need of communication with the affected participants. Instead of broadcast messages to all shareholders, only a comparatively small message must be sent to the combiner via a secure channel.

As an example, we gave an efficient implementation based on Shamir's threshold scheme which is ideal from a shareholder's point of view.

We mention that the described ideas allow the implementation of a scheme that realizes general non-monotone access structures. These thoughts are not limited to single secret sharing schemes but are easily transferable to multi secret sharing schemes.

Acknowledgements

The author thanks Keith Martin and the anonymous reviewers for their helpful observations and comments.

References

1. Barwick, S.G., Jackson, W.-A., Martin, K.M., O'Keefe, C.: Optimal updating of ideal threshold schemes. Australasian Journal of Combinatorics 36, 123–132 (2006)
2. Barwick, S.G., Jackson, W.-A., Martin, K.M.: Updating the parameters of a threshold scheme by minimal broadcast. IEEE Transactions on Information Theory 51(2), 620–633 (2005)
3. Blakley, G.R.: Safeguarding Cryptographic Keys. In:Proceesings AFIPS 1979, National Computer Conference, vol. 48, pp. 313–317 (1979)
4. Blakley, B., Blakley, G.R., Chan, A.H., Massey, J.L.: Threshold schemes with disenrollment. In: Brickell, E.F. (ed.) CRYPTO 1992. LNCS, vol. 740, pp. 540–548. Springer, Heidelberg (1993)
5. Blundo, C., Cresti, A., De Santis, A., Vaccaro, U.: Fully dynamic secret sharing schemes. In: Stinson, D.R. (ed.) CRYPTO 1993. LNCS, vol. 773, pp. 110–125. Springer, Heidelberg (1994)
6. Cachin, C.: On-Line Secret Sharing. In: Boyd, C. (ed.) Cryptography and Coding 1995. LNCS, vol. 1025, pp. 190–198. Springer, Heidelberg (1995)

7. Ghodosi, H., Pieprzyk, J., Safavi-Naini, R.: Dynamic Threshold Cryptosystems: A New Scheme in Group Oriented Cryptography. In: Proceedings of PRAGOCRYPT 1996. International Conference on the Theory and Applications of Cryptology, pp. 370–379 (1996)
8. Martin, K.M.: Untrustworthy Participants in Secret Sharing Schemes, In: Cryptography and Coding III, pp. 255–264. Oxford University Press, Oxford (1993)
9. Martin, K.M.: Dynamic access policies for unconditionally secure secret sharing schemes. In: Proceedings of IEEE Information Theory Workshop (ITW 2005), Awaji Island, Japan (2005)
10. Pedersen, T.: A Threshold Cryptosystem without a Trusted Party. In: Davies, D.W. (ed.) EUROCRYPT 1991. LNCS, vol. 547, pp. 522–526. Springer, Heidelberg (1991)
11. Shamir, A.: How to share a secret. Communications of the ACM 22(11), 612–613 (1979)
12. Simmons, G.J.: How to (really) share a secret. In: Goldwasser, S. (ed.) CRYPTO 1988. LNCS, vol. 403, pp. 390–448. Springer, Heidelberg (1988)
13. Simmons, G.J.: Prepositioned shared secret and/or shared control schemes. In: Quisquater, J.-J., Vandewalle, J. (eds.) EUROCRYPT 1989. LNCS, vol. 434, pp. 436–467. Springer, Heidelberg (1990)
14. Simmons, G.J.: An introduction to shared secret and/or shared control schemes and their applications. In: Contemporary Cryptology, pp. 441–497. IEEE Press, Los Alamitos (1992)
15. Stinson, D.R.: An explication of secret sharing schemes. In: Designs, Codes and Cryptography, pp. 357–390. Kluwer Academic Publishers, Dordrecht (1992)
16. Stoer, J., Bulirsch, R.: Introduction to Numerical Analysis. Springer, Heidelberg (2002)
17. Tompa, M., Woll, H.: How To Share a Secret with Cheaters. Journal of Cryptology, 133–138 (1988)

New Attacks on the Stream Cipher TPy6 and Design of New Ciphers the TPy6-A and the TPy6-B[*]

Gautham Sekar, Souradyuti Paul, and Bart Preneel

Katholieke Universiteit Leuven, Dept. ESAT/COSIC,
Kasteelpark Arenberg 10,
B–3001, Leuven-Heverlee, Belgium
{gautham.sekar,souradyuti.paul,bart.preneel}@esat.kuleuven.be

Abstract. The stream ciphers Py, Pypy and Py6 were designed by Biham and Seberry for the ECRYPT-eSTREAM project in 2005. The ciphers were promoted to the 'Focus' ciphers of the Phase II of the eS-TREAM project. However, due to some cryptanalytic results, strengthened versions of the ciphers, namely, the TPy, the TPypy and the TPy6 were built. In this paper, we find hitherto unknown weaknesses in the keystream generation algorithms of the Py6 and its stronger variant the TPy6. Exploiting these weaknesses, a large number of distinguishing attacks are mounted on the ciphers, the best of which works with $2^{224.6}$ data and comparable time. In the second part, we present two new ciphers derived from the TPy6, namely, the TPy6-A and the TPy6-B, whose performances are 2.65 cycles/byte and 4.4 cycles/byte on Pentium III. As a result, to the best of our knowledge, on Pentium platforms the TPy6-A becomes the fastest stream cipher in the literature. Based on our security analysis, we conjecture that no attacks lower than the brute force are possible on the ciphers TPy6-A and TPy6-B.

1 Introduction

At first, we recall a brief history of the Py-family of ciphers.

Timeline: the Py-family of Ciphers

- **April 2005.** The ciphers Py and Py6, designed by Biham and Seberry, were submitted to the ECRYPT project for analysis and evaluation in the category of software based stream ciphers [2]. The impressive speed of the cipher Py in software (about 2.5 times faster than the RC4) made it one of the fastest and most attractive contestants.

---

[*] This work was supported in part by the Concerted Research Action (GOA) Ambiorics 2005/11 of the Flemish Government, by the IAP Programme P6/26 BCRYPT of the Belgian State (Belgian Science Policy), and in part by the European Commission through the IST Programme under Contract IST-2002-507932 ECRYPT. The first and the second authors are supported by IWT SoBeNeT project and an IBBT (Interdisciplinary Institute for Broadband Technology) project respectively.

- **March 2006 (at FSE 2006).** Paul, Preneel and Sekar reported distinguishing attacks with $2^{89.2}$ data and comparable time against the cipher Py [7]. Crowley [4] later reduced the complexity to 2^{72} by employing a Hidden Markov Model.
- **March 2006 (at the Rump session of FSE 2006).** A new cipher, namely Pypy, was proposed by the designers to rule out the aforementioned distinguishing attacks on Py [3].
- **May 2006 (presented at Asiacrypt 2006).** Distinguishing attacks were reported against Py6 with 2^{68} data and comparable time by Paul and Preneel [8].
- **October 2006 (presented at Eurocrypt 2007).** Wu and Preneel showed key recovery attacks against the ciphers Py, Pypy, Py6 with chosen IVs. This attack was subsequently improved by Isobe *et al.* [6].
- **January 2007.** Three new ciphers TPypy, TPy, TPy6 were proposed by the designers [1]; the ciphers can very well be viewed as the strengthened versions of the previous ciphers Py, Pypy and Py6.
- **February 2007.** Sekar *et al.* published attacks on TPy and TPypy, each of which requires 2^{281} data and comparable time [9].
- **August 2007 (presented at SAC 2007).** Tsunoo *et al.* showed a distinguishing attack on TPypy with a data complexity of 2^{199} [13].
- **October 2007 (presented at ISC 2007).** Sekar *et al.* showed attacks on TPy, the best of which requires $2^{268.6}$ data and comparable time [10].
- **December 2007 (presented at Indocrypt 2007).** Sekar *et al.* showed related-key attacks on the Py, the Pypy, the TPy and the TPypy, each requiring 2^{192} data and comparable time [11].

In this paper, we detect new bias-inducing internal states of the TPy6 to build distinguishing attacks on the cipher. The bias-inducing states are similar in spirit to those of the other distinguishing attacks on the Py-family [8,10], however, they were not known so far. The characterization of the bias-inducing states is given in full detail in Theorem 1.

We also design two new ciphers the TPy6-A and the TPy6-B by changing the variable rotations of the keystream generation of the TPy6 to constant rotations. This simple change makes the new ciphers (1) operate with fewer instructions and (2) conjecturally more secure than the TPy6.

2   Notation and Convention

Algorithm 1 describes the keystream generation part of the TPy6 and the Py6. The notation we followed is described below.

- The mth bit ($m = 0$ denotes the least significant bit or lsb) of the first output-word generated at round n is denoted by $O_{n(m)}$. The second output-word is not used anywhere in our analysis.

- P_n, Y_{n+1} and s_n are the inputs to the algorithm at round n. When this convention is followed, we see that $O_n = (ROTL32(s_n, 25) \oplus Y_n[64]) + Y_n[P_n[26]]$- the index '$n$' is maintained throughout the expression.
- The $ROTL32(x, k)$ denotes that the variable x is cyclically rotated to the left by k bits.
- $Y_n[m]$, $P_n[m]$ denote the mth elements of array Y_n and P_n respectively.
- $Y_n[m]_i$, $P_n[m]_i$ denote the ith bit ($i = 0$ denotes the lsb) of $Y_n[m]$, $P_n[m]$ respectively.
- The symbol '&' denotes the *and* operator.
- The operators '+' and '−' denote *addition modulo* 2^{32} and *subtraction modulo* 2^{32} respectively, except when used with expressions which relate two elements of array P. In this case they denote *addition and subtraction over* $\mathbb{Z}$.
- The symbol '$\oplus$' denotes bitwise *exclusive-or* and $\bigcap$ denotes set intersection.
- In $O_{n(i)}$, $s_{n(i)}$ and $Y_n[P_m[X]]_i$, the index representing bit position, i.e., i denotes $i \bmod 32$.
- $Y_n^c[P_m[X]]_i$ denotes the complement of $Y_n[P_m[X]]_i$.
- The pseudorandom bit generation algorithm of a stream cipher is denoted by PRBG.

Algorithm 1. The keystream generation algorithm of the TPy6 (and the Py6)

Require: $Y[-3, ..., 64]$, $P[0, ..., 63]$, a 32-bit variable s
Ensure: 64-bit random output
    /*Update and rotate P*/
1. swap $(P[0], P[Y[43]\&63])$;
2. rotate (P);
    /* Update s*/
3. $s+ = Y[P[18]] − Y[P[57]]$;
4. $s = ROTL32(s, ((P[26] + 18)\&31))$;
    /* Output 4 or 8 bytes (least significant byte first)*/
5. output $((ROTL32(s, 25) \oplus Y[64]) + Y[P[8]])$;
6. output $((\quad\quad s\quad\quad \oplus Y[-1]) + Y[P[21]])$;
    /* Update and rotate Y*/
7. $Y[-3] = (ROTL32(s, 14) \oplus Y[-3]) + Y[P[48]]$;
8. rotate(Y);

---

3   Distinguishing Attacks on the Py6 and the TPy6

We detect a large number of input-output correlations of TPy6 and Py6 that allow us to build distinguishers. The weak states which lead to the best distinguishing attack are outlined in the following theorem.

Theorem 1. $O_{1(i)} \oplus O_{3(i+7)} \oplus O_{7(i+7)} \oplus O_{8(i+7)} = 0$ *if the following 17 conditions are simultaneously satisfied.*

1. $P_1[26] \equiv -18 \bmod 32$ *(event E_1)*, 2. $P_2[26] \equiv 7 \bmod 32$ *(event E_2)*, 3. $P_3[26] \equiv -4 \bmod 32$ *(event E_3)*, 4. $P_7[26] \equiv 3 \bmod 32$ *(event E_4)*, 5. $P_8[26] \equiv 3 \bmod 32$ *(event E_5)*, 6. $P_1[18] = P_2[57] + 1$ *(event E_6)*, 7. $P_1[57] = P_2[18] + 1$ *(event E_7)*, 8. $P_7[18] = P_8[18] + 1$ *(event E_8)*, 9. $P_7[57] = P_8[57] + 1$ *(event E_9)*, 10. $P_3[18] = 62$ *(event E_{10})*, 11. $P_1[8] = P_3[57] + 2$ *(event E_{11})*, 12. $P_1[18] = 3$ *(event E_{12})*, 13. $P_3[8] = 0$ *(event E_{13})*, 14. $P_1[57] = P_7[8] + 6$ *(event E_{14})*, 15. $P_7[48] = 60$ *(event E_{15})*, 16. $P_6[48] = P_8[8] + 2$ *(event E_{16})*, 17. $d_{7(i-7)} \oplus d_{8(i-7)} \oplus c_{1(i)} \oplus d_{3(i)} \oplus d_{1(i+7)} \oplus c_{3(i+7)} \oplus c_{7(i+7)} \oplus e_{7(i+7)} \oplus c_{8(i+7)} \oplus e_{8(i+7)} = 0.$[1]

Proof. First, we state and prove two lemmata which will be used to establish the theorem.

Lemma 1. *If*

1. $P_1[26] \equiv -18 \bmod 32$, 2. $P_3[26] \equiv -4 \bmod 32$, 3. $P_7[26] \equiv 3 \bmod 32$, 4. $P_8[26] \equiv 3 \bmod 32$

then the following equations are satisfied:

1. $O_{1(i)} = s_{0(i+7)} \oplus Y_1[P_1[18]]_{i+7} \oplus Y_1^c[P_1[57]]_{i+7} \oplus Y_1[256]_i \oplus Y_1[P_1[8]]_i$
   $\oplus c_{1(i)} \oplus d_{1(i+7)},$
2. $O_{3(i+7)} = s_{2(i)} \oplus Y_3[P_3[18]]_i \oplus Y_3^c[P_3[57]]_i \oplus Y_3[256]_{i+7} \oplus Y_3[P_3[8]]_{i+7}$
   $\oplus c_{3(i+7)} \oplus d_{3(i)},$
3. $O_{7(i+7)} = Y_7[P_7[18]]_{i-7} \oplus Y_7^c[P_7[57]]_{i-7} \oplus Y_6[-3]_{i+7} \oplus Y_7[P_7[8]]_{i+7}$
   $\oplus Y_6[P_6[48]]_{i+7} \oplus c_{7(i+7)} \oplus d_{7(i-7)} \oplus e_{7(i+7)},$
4. $O_{8(i+7)} = Y_8[P_8[18]]_{i-7} \oplus Y_8^c[P_8[57]]_{i-7} \oplus Y_7[-3]_{i+7} \oplus Y_8[P_8[8]]_{i+7}$
   $\oplus Y_7[P_7[48]]_{i+7} \oplus c_{8(i+7)} \oplus d_{8(i-7)} \oplus e_{8(i+7)}.$

Proof. From Figure 1, we get

$$Y_n[i] = Y_{n+1}[i-1] \tag{1}$$

when $-2 \leq i \leq 64$. When $i = -3$,

$$Y_{n+1}[64] = (ROTL32(s_i, 14) \oplus Y_n[-3]) + Y_n[P_n[48]].$$

Generalizing (1), we have

$$Y_n[i] = Y_{n+k}[i-k] \tag{2}$$

when $-3 \leq i - k \leq 63$. Line 5 of Algorithm 1 gives

$$O_7 = (ROTL32(s_7, 25) \oplus Y_7[64]) + Y_7[P_7[8]]. \tag{3}$$

---

[1] The terms c, d, e are the carries generated in certain expressions, the descriptions of which can be found in the proof of Theorem 1.

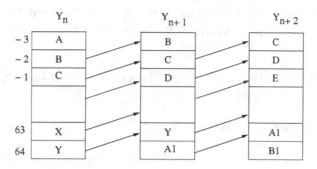

Fig. 1. The figure shows the update of the S-box Y. $Y_n[i] = Y_{n+1}[i-1]$ when $-2 \leq i \leq 64$. $Y_{n+1}[64] = A1$ when $i = -3$ and $A1 = (ROTL32(s_n, 14) \oplus A) + Y_n[P_n[48]]$. Generalizing the above, we can write $Y_n[i] = Y_{n+k}[i-k]$ when $-3 \leq i - k \leq 63$.

Let the c_7 denote the carry in the above equation. Since $ROTL32(s_7, 25)_i = s_{7(i-25 \bmod 32)}$,

$$O_{7(i)} = s_{7(i-25 \bmod 32)} \oplus Y_7[64]_i \oplus Y_7[P_7[8]]_i \oplus c_{7(i)}. \tag{4}$$

Lines 3 and 4 of Algorithm 1 give us

$$s_7 = ROTL32(s_6 + Y_7[P_7[18]] - Y_7[P_7[57]], P_7[26] + 18 \bmod 32) \tag{5}$$

$$\Rightarrow s_{7(j)} = s_{6(j-k \bmod 32)} \oplus Y_7[P_7[18]]_{j-k \bmod 32} \oplus Y_7^c[P_7[57]]_{j-k \bmod 32}$$
$$\oplus d_{7(j-k \bmod 32)} \tag{6}$$

where $k = P_7[26] + 18 \bmod 32$, $d_{7(i)} = f_{7(i)} \oplus g_{7(i)}$ and $d_{7(0)} = 1$ (f_7 and g_7 are the carry terms in (5) which are explained in Sect. 4.2). For simplicity, henceforth we denote $X_{(i \bmod 32)}$ by $X_{(i)}$. Thus (6) becomes,

$$s_{7(j)} = s_{6(j-k)} \oplus Y_7[P_7[18]]_{j-k} \oplus Y_7^c[P_7[57]]_{j-k} \oplus d_{7(j-k)}. \tag{7}$$

If $j = i - 25 \bmod 32$, then (7) becomes

$$s_{7(i-25)} = s_{6(i-k-25)} \oplus Y_7[P_7[18]]_{i-k-25} \oplus Y_7^c[P_7[57]]_{i-k-25} \oplus d_{7(i-k-25)}. \tag{8}$$

Substituting (8) in (4), we get,

$$O_{7(i)} = s_{6(i-k-25)} \oplus Y_7[P_7[18]]_{i-k-25} \oplus Y_7^c[P_7[57]]_{i-k-25} \oplus Y_7[64]_i$$
$$\oplus Y_7[P_7[8]]_i \oplus c_{7(i)} \oplus d_{7(i-k-25)}. \tag{9}$$

Next, we have

$$Y_7[64] = (ROTL32(s_6, 14) \oplus Y_6[-3]) + Y_6[P_6[48]], \tag{10}$$

$$Y_7[64]_i = s_{6(i-14)} \oplus Y_6[-3]_i \oplus Y_6[P_6[48]]_i \oplus e_{7(i)} \tag{11}$$

where e_7 is the carry term in (10). Substituting (11) in (9), we get,

$$O_{7(i)} = s_{6(i-k-25)} \oplus s_{6(i-14)} \oplus Y_7[P_7[18]]_{i-k-25} \oplus Y_7^c[P_7[57]]_{i-k-25} \oplus Y_6[-3]_i$$
$$\oplus Y_7[P_7[8]]_i \oplus Y_6[P_6[48]]_i \oplus c_{7(i)} \oplus d_{7(i-k-25)} \oplus e_{7(i)}. \tag{12}$$

Now, if $k = -11$ (i.e., $k \equiv -11 \bmod 32 \Rightarrow P_7[26]+18 \equiv -11 \bmod 32 \Rightarrow P_7[26] \equiv 3 \bmod 32$) then $s_{6(i-k-25)} \oplus s_{6(i-14)} = 0$. Hence, when $P_7[26] \equiv 3 \bmod 32$, (12) becomes

$$O_{7(i)} = Y_7[P_7[18]]_{i-14} \oplus Y_7^c[P_7[57]]_{i-14} \oplus Y_6[-3]_i \oplus Y_7[P_7[8]]_i$$
$$\oplus Y_6[P_6[48]]_i \oplus c_{7(i)} \oplus d_{7(i-14)} \oplus e_{7(i)}. \tag{13}$$

By similar arguments, when $P_8[26] \equiv 3 \bmod 32$,

$$O_{8(i)} = Y_8[P_8[18]]_{i-14} \oplus Y_8^c[P_8[57]]_{i-14} \oplus Y_7[-3]_i \oplus Y_8[P_8[8]]_i$$
$$\oplus Y_7[P_7[48]]_i \oplus c_{8(i)} \oplus d_{8(i-14)} \oplus e_{8(i)}. \tag{14}$$

From (9), we get

$$O_{1(i)} = s_{0(i-k-25)} \oplus Y_1[P_1[18]]_{i-k-25} \oplus Y_1^c[P_1[57]]_{i-k-25} \oplus Y_1[64]_i$$
$$\oplus Y_1[P_1[8]]_i \oplus c_{1(i)} \oplus d_{1(i-k-25)}. \tag{15}$$

When $k = 0$ (i.e., $P_1[26] \equiv -18 \bmod 32$), the above equation reduces to

$$O_{1(i)} = s_{0(i+7)} \oplus Y_1[P_1[18]]_{i+7} \oplus Y_1^c[P_1[57]]_{i+7} \oplus Y_1[64]_i \oplus Y_1[P_1[8]]_i$$
$$\oplus c_{1(i)} \oplus d_{1(i+7)}. \tag{16}$$

Similarly, when $P_3[26] \equiv -4 \bmod 32$, we have

$$O_{3(i+7)} = s_{2(i)} \oplus Y_3[P_3[18]]_i \oplus Y_3^c[P_3[57]]_i \oplus Y_3[64]_{i+7} \oplus Y_3[P_3[8]]_{i+7}$$
$$\oplus c_{3(i+7)} \oplus d_{3(i)}. \tag{17}$$

From (13) and (14), we derive the following results:

$$O_{7(i+7)} = Y_7[P_7[18]]_{i-7} \oplus Y_7^c[P_7[57]]_{i-7} \oplus Y_6[-3]_{i+7} \oplus Y_7[P_7[8]]_{i+7}$$
$$\oplus Y_6[P_6[48]]_{i+7} \oplus c_{7(i+7)} \oplus d_{7(i-7)} \oplus e_{7(i+7)}, \tag{18}$$
$$O_{8(i+7)} = Y_8[P_8[18]]_{i-7} \oplus Y_8^c[P_8[57]]_{i-7} \oplus Y_7[-3]_{i+7} \oplus Y_8[P_8[8]]_{i+7}$$
$$\oplus Y_7[P_7[48]]_{i+7} \oplus c_{8(i+7)} \oplus d_{8(i-7)} \oplus e_{8(i+7)}. \tag{19}$$

This completes the proof.   □

Now we state the second lemma.

Lemma 2. $s_{0(i+7)} = s_{2(i)}$ *if the following conditions are simultaneously satisfied,*

1. $P_1[26] \equiv -18 \bmod 32$,
2. $P_2[26] \equiv 7 \bmod 32$,
3. $P_1[18] = P_2[57] + 1$,
4. $P_1[57] = P_2[18] + 1$.

Table 1. Terms generated in $O_{1(i)} \oplus O_{3(i+7)} \oplus O_{7(i+7)} \oplus O_{8(i+7)}$, given the events E_1 to E_7 simultaneously occur (the terms are grouped by their bit positions)

Bit position: $i-7$	Bit position: i	Bit position: $i+7$
$Y_7[P_7[18]]$	$Y_1[64]$	$Y_1[P_1[18]]$
$Y_7[P_7[57]]$	$Y_1[P_1[8]]$	$Y_1[P_1[57]]$
$Y_8[P_8[18]]$	$Y_3[P_3[18]]$	$Y_3[256]$
$Y_8[P_8[57]]$	$Y_3[P_3[57]]$	$Y_3[P_3[8]]$
Carries	Carries	$Y_6[P_6[48]]$
		$Y_6[-3]$
		$Y_7[P_7[8]]$
		$Y_7[P_7[48]]$
		$Y_7[-3]$
		$Y_8[P_8[8]]$
		Carries

Proof. Equation (5) gives us:

$$s_1 = ROTL32(s_0 + Y_1[P_1[18]] - Y_1[P_1[57]], P_1[26] + 18 \bmod 32).$$

The first condition ($P_1[26] \equiv -18 \bmod 32$) reduces this to

$$s_1 = s_0 + Y_1[P_1[18]] - Y_1[P_1[57]].$$

Therefore,

$$s_2 = ROTL32(s_0 + Y_2[P_2[18]] - Y_2[P_2[57]] + Y_1[P_1[18]] - Y_1[P_1[57]],$$
$$P_2[26] + 18 \bmod 32).$$

Conditions 3 and 4, when used with (1), reduce the above equation to

$$s_2 = ROTL32(s_0, P_2[26] + 18 \bmod 32).$$

Finally, with condition 2 (i.e., $P_2[26] \equiv 7 \bmod 32$), the previous equation becomes

$$s_2 = ROTL32(s_0, 25)$$
$$\Rightarrow s_{2(i)} = ROTL32(s_0, 25)_i = s_{0(i-25)}$$
$$= s_{0(i+7)}. \tag{20}$$

This completes the proof. $\qquad\qquad\qquad\qquad\qquad\qquad\qquad\qquad\qquad\qquad\qquad\square$

Now we observe that, when the conditions listed under (i) Lemma 1 (i.e., events E_1, E_3, E_4 and E_5) and (ii) Lemma 2 (i.e., events E_1, E_2, E_6 and E_7) are simultaneously satisfied, then the expression $O_{1(i)} \oplus O_{3(i+7)} \oplus O_{7(i+7)} \oplus O_{8(i+7)}$ is the XOR of the terms which are listed in Table 1 (grouped according to the bit positions).[2] Similarly, the 'carries' in Table 1 are elaborated in Table 2.

---

[2] Note that none of the terms listed in Table 1 is of the form A^c because we used the fact that $A^c \oplus B^c = A \oplus B$ in (16), (17), (18) and (19).

Table 2. Carry terms generated in $O_{1(i)} \oplus O_{3(i+7)} \oplus O_{7(i+7)} \oplus O_{8(i+7)}$ grouped by their bit positions

Bit position: $i-7$	Bit position: i	Bit position: $i+7$
d_7	c_1	d_1
d_8	d_3	c_3
		c_7
		e_7
		c_8
		e_8

If the Y-terms in Table 1 are pairwise equated (this is achieved when the events E_8 through to E_{16} occur) then we get

$$O_{1(i)} \oplus O_{3(i+7)} \oplus O_{7(i+7)} \oplus O_{8(i+7)} = d_{7(i-7)} \oplus d_{8(i-7)} \oplus c_{1(i)} \oplus d_{3(i)} \oplus d_{1(i+7)}$$
$$\oplus c_{3(i+7)} \oplus c_{7(i+7)} \oplus e_{7(i+7)} \oplus c_{8(i+7)}$$
$$\oplus e_{8(i+7)}. \tag{21}$$

Now, when the RHS of (21) equals zero (i.e., E_{17} occurs) we get

$$O_{1(i)} \oplus O_{3(i+7)} \oplus O_{7(i+7)} \oplus O_{8(i+7)} = 0.$$

This completes the proof of Theorem 1. □

4   Computation of the Bias

In this section, we quantify the bias in the outputs of TPy6 (and hence Py6) induced by the fortuitous events similar to the one described in Sect. 3. Now it is important to note that there may be *more than one set of 17 conditions* possible, where each of them results in $O_{1(i)} \oplus O_{3(i+7)} \oplus O_{7(i+7)} \oplus O_{8(i+7)} = 0$ (let us assume that there are n such sets). In Theorem 1, we listed one such set. Our experiments suggest that these n sets are mutually independent, however, a formal proof of that is nontrivial.

Each of the events E_1 to E_5 occurs with approximate probability $\frac{1}{32}$ and each of the events E_6 to E_{16} occurs with probability which is approximately $\frac{1}{64}$. Let p denote the probability that condition 17 is satisfied. Let F denote the event $\bigcap_{j=1}^{16} E_j$. Therefore,

$$P[F] = \left(\frac{1}{32}\right)^5 \cdot \left(\frac{1}{64}\right)^{11}.$$

We see that there are n F-like events (i.e., the intersection of 16 conditions). Let F_n denote the union of these n events. Since, each event occurs with approximately the same probability,

$$P[F_n] \approx n \cdot P[F]$$
$$\approx n \cdot (\frac{1}{32})^5 \cdot (\frac{1}{64})^{11}$$
$$= n \cdot \frac{1}{2^{91}}.$$

From Table 1, we get the maximum number of ways that terms of a particular column can be pairwise equated and hence the upper bound on n can be calculated to be $2 \cdot 2 \cdot 945 = 3780$, that is $n < 3780$.

4.1 Formulating the Bias

Now, we establish a formula to compute $P[O_{1(i)} \oplus O_{3(i+7)} \oplus O_{7(i+7)} \oplus O_{8(i+7)} = 0]$, under the assumption of a perfectly random key/IV setup and the uniformity of bits when F_n does not occur. Our experiments suggest that it is infeasible to find a set of conditions such that the overall bias (computed on the basis of the aforementioned assumption of randomness in the event that F_n does not occur) is canceled or reduced in magnitude. Therefore, this assumption is reasonable. Let T denote $O_{1(i)} \oplus O_{3(i+7)} \oplus O_{7(i+7)} \oplus O_{8(i+7)}$. Then using Bayes' rule we get

$$P[T = 0] = P[T = 0|F_n \cap E_{17}] \cdot P[F_n \cap E_{17}] + P[T = 0|F_n^c \cup E_{17}^c] \cdot P[F_n^c \cup E_{17}^c]$$
$$= P[T = 0|F_n \cap E_{17}] \cdot P[F_n \cap E_{17}] + P[T = 0|F_n^c \cap E_{17}] \cdot P[F_n^c \cap E_{17}]$$
$$+ P[T = 0|F_n \cap E_{17}^c] \cdot P[F_n \cap E_{17}^c] + P[T = 0|F_n^c \cap E_{17}^c] \cdot P[F_n^c \cap E_{17}^c]$$
$$= 1 \cdot (n \cdot p \cdot \frac{1}{2^{91}}) + \frac{1}{2} \cdot (1 - n \cdot \frac{1}{2^{91}}) \cdot p + 0 \cdot P[F_n \cap E_{17}^c]$$
$$+ \frac{1}{2} \cdot (1 - n \cdot \frac{1}{2^{91}}) \cdot (1 - p)$$
$$= \frac{1}{2} + n \cdot (2p - 1) \cdot \frac{1}{2^{92}}. \tag{22}$$

Hence, we see that the distribution of the outputs $(O_{1(i)}, O_{3(i+7)}, O_{7(i+7)}, O_{8(i+7)})$ is biased. The bias is equal to $n \cdot (2p - 1) \cdot \frac{1}{2^{92}}$. In the following section, we provide formulas to compute p, i.e., the probability that E_{17} occurs; or more generally, the probability that the 17th condition of each of the n F-like events occurs, i.e., $P[d_{7(i-7)} \oplus d_{8(i-7)} \oplus c_{1(i)} \oplus d_{3(i)} \oplus d_{1(i+7)} \oplus c_{3(i+7)} \oplus c_{7(i+7)} \oplus e_{7(i+7)} \oplus c_{8(i+7)} \oplus e_{8(i+7)}] = 0$.

4.2 Biases in the Carry Terms

In this section, we provide formulas to calculate the bias in the carry terms. The carry terms c and e are generated in expressions of the form $(S \oplus X) + Z$. We now proceed to calculate $P[c_{l(i)} = 0]$ assuming that S, X and Z are uniformly distributed and independent. Under this assumption, $P[S_i = 0] = P[X_i = 0] = P[Z_i = 0] = \frac{1}{2}$, that is, the probability that the carry bit at position i equals zero depends only on i. Stated otherwise, $P[c_{(i)} = 0] = P[e_{(i)} = 0]$. Let $P[c_{(i)} = 0]$ be denoted by p_i. Since there is no carry on the lsb, $p_0 = 1$.

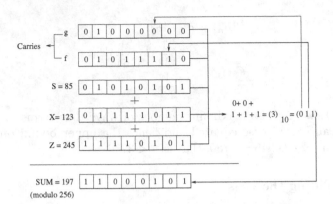

Fig. 2. An example showing how the carries are generated when three 8-bit variables $S = 85$, $X = 123$ and $Z = 245$ are added

Using Bayes' rule we get

$$p_i = \frac{p_{i-1}}{2} + \frac{1}{4}.$$

Details of the calculation are provided in the full version [12]. Solving this recursion, given $p_0 = 1$, we get

$$p_i = \frac{1}{2} + \frac{1}{2^{i+1}}. \tag{23}$$

Now, the carry terms f and g are generated in expressions of the form $S+X-Z$. This can be rewritten as $S+X+Z^c+1$ since the additions in these two expressions are modulo 2^{32}. The presence of two carries in $S + X + Z$ is demonstrated using the Figure 2. The carries generated in $S + X + Z^c + 1$ can be thought of as carries generated in $S+X+A$ where $A = Z^c$ and the carries on the lsb $f_{(0)} = 1$, $g_{(0)} = 0$. Let q_i denote $P[f_{(i)} = 0]$ and r_i denote $P[g_{(i)} = 0]$. Hence, $q_0 = 0$, $r_0 = 1$ and $r_1 = 1$. Now we have Table 3.

From Table 3, using Bayes' rule we get

$$q_i = \frac{1}{2} + \frac{5 \cdot q_{i-1} \cdot r_{i-1}}{8} - \frac{q_{i-1}}{4} - \frac{r_{i-1}}{4}, \tag{24}$$

$$r_{i+1} = \frac{1}{2} - \frac{q_{i-1} \cdot r_{i-1}}{4} + \frac{3 \cdot q_{i-1}}{8} + \frac{3 \cdot r_{i-1}}{8}. \tag{25}$$

Using the initial conditions, $q_0 = 0$, $r_0 = 1$ and $r_1 = 1$, q_i and r_i are computed recursively. Since $d_{m(i)}$ denotes $f_{m(i)} \oplus g_{m(i)}$ for any $m > 0$,

1. $P[d_{7(i-7)} = 0] = P[d_{8(i-7)} = 0] = q_{i-7 \bmod 32} \cdot r_{i-7 \bmod 32}$
   $+ (1 - q_{i-7 \bmod 32}) \cdot (1 - r_{i-7 \bmod 32})$,
2. $P[c_{1(i)} = 0] = \frac{1}{2} + \frac{1}{2^{i+1}}$,
3. $P[d_{3(i)} = 0] = q_i \cdot r_i + (1 - q_i) \cdot (1 - r_i)$,

Table 3. Truth table for computing q_i and r_{i+1} using q_{i-1} and r_{i-1} (NR=Not Required)

$f_{(i-1)}$	$g_{(i-1)}$	$S_{(i-1)}$	$X_{(i-1)}$	$Z_{(i-1)}$	$f_{(i)}$	$g_{(i+1)}$	Probability
0	0	0	0	0	0	0	$\frac{q_{i-1}\cdot r_{i-1}}{8}$
0	0	0	0	1	0	0	$\frac{q_{i-1}\cdot r_{i-1}}{8}$
0	0	0	1	0	0	0	$\frac{q_{i-1}\cdot r_{i-1}}{8}$
0	0	0	1	1	1	0	NR
0	0	1	0	0	0	0	$\frac{q_{i-1}\cdot r_{i-1}}{8}$
0	0	1	0	1	1	0	NR
0	0	1	1	0	1	0	NR
0	0	1	1	1	0	0	$\frac{q_{i-1}\cdot r_{i-1}}{8}$
0	1	0	0	0	0	0	$\frac{q_{i-1}\cdot(1-r_{i-1})}{8}$
0	1	0	0	1	1	0	NR
0	1	0	1	0	1	0	NR
0	1	0	1	1	1	0	NR
0	1	1	0	0	1	0	NR
0	1	1	0	1	1	0	NR
0	1	1	1	0	1	0	NR
0	1	1	1	1	0	1	$\frac{q_{i-1}\cdot(1-r_{i-1})}{8}$
1	0	0	0	0	0	0	$\frac{(1-q_{i-1})\cdot r_{i-1}}{8}$
1	0	0	0	1	1	0	NR
1	0	0	1	0	1	0	NR
1	0	0	1	1	1	0	NR
1	0	1	0	0	1	0	NR
1	0	1	0	1	1	0	NR
1	0	1	1	0	1	0	NR
1	0	1	1	1	0	1	$\frac{(1-q_{i-1})\cdot r_{i-1}}{8}$
1	1	0	0	0	1	0	NR
1	1	0	0	1	1	0	NR
1	1	0	1	0	1	0	NR
1	1	0	1	1	0	1	$\frac{(1-q_{i-1})\cdot(1-r_{i-1})}{8}$
1	1	1	0	0	1	0	NR
1	1	1	0	1	0	1	$\frac{(1-q_{i-1})\cdot(1-r_{i-1})}{8}$
1	1	1	1	0	0	1	$\frac{(1-q_{i-1})\cdot(1-r_{i-1})}{8}$
1	1	1	1	1	0	1	$\frac{(1-q_{i-1})\cdot(1-r_{i-1})}{8}$

4. $P[d_{1(i+7)} = 0] = q_{i+7 \bmod 32} \cdot r_{i+7 \bmod 32}$
   $+ (1 - q_{i+7 \bmod 32}) \cdot (1 - r_{i+7 \bmod 32})$,

5. $P[c_{3(i+7)} = 0] = P[c_{7(i+7)} = 0] = P[e_{7(i+7)} = 0] = P[c_{8(i+7)} = 0]$
   $= P[e_{8(i+7)} = 0] = \frac{1}{2} + \frac{1}{2^{(i+7 \bmod 32)+1}}$.

Using the above formulas, the value of p can be computed for any given i. Running simulation, we find that the maximum bias in the chosen outputs occurs when $i = 25$ which corresponds to $p = 0.5 - 2^{-34.2}$. Hence, (22) gives us

$$P[T = 0] = \frac{1}{2} - \frac{n}{2^{125.2}}$$

$$\Rightarrow P[T = 1] = \frac{1}{2} + \frac{n}{2^{125.2}},$$

when $i = 25$. Substituting $n = 3780$ in the above equation, we get:

$$P[T = 1] = \frac{1}{2} + \frac{1}{2^{113.3}}. \tag{26}$$

This is an upper bound on the probability that the outputs $(O_{1(i)}, O_{3(i+7)}, O_{7(i+7)}, O_{8(i+7)})$ of TPy6 (and hence Py6) are biased. From Sect. 3, we found that $n \geq 1$. From the previous discussion, we see that $n < 3780$. Hence, $1 \leq n < 3780$. If $n = 1$, then $P[T = 1] = \frac{1}{2} + \frac{1}{2^{125.2}}$. Thus,

$$\frac{1}{2}(1 + \frac{1}{2^{124.2}}) \leq P[T = 1] < \frac{1}{2}(1 + \frac{1}{2^{112.3}}). \tag{27}$$

5   The Distinguisher

A distinguisher is an algorithm which distinguishes a given stream of bits from a stream of bits generated by a perfect PRBG. The distinguisher is constructed by collecting sufficiently many outputs $(O_{1(25)}, O_{3(0)}, O_{7(0)}, O_{8(0)})$ generated by as many key/IVs. To compute the minimum number of samples required to establish the distinguisher, we use the following corollary of a theorem from [5].

Corollary 1. *If an event e occurs in a distribution X with probability p and in Y with probability $p(1 + q)$ then, if $p = \frac{1}{2}$, $O(\frac{1}{q^2})$ samples are required to distinguish X from Y with non-negligible probability of success.*

In the present case, e is the event $O_{1(25)} \oplus O_{3(0)} \oplus O_{7(0)} \oplus O_{8(0)} = 0$, X is the distribution of the outputs O_1, O_3, O_7 and O_8 produced by a perfectly random keystream generator and Y is the distribution of the outputs produced by TPy6. From (27), $p = \frac{1}{2}$ and the highest value of $q = \frac{1}{2^{112.3}}$. Hence $O(\frac{1}{(2^{-112.3})^2}) = O(2^{224.6})$ output samples are needed to construct the best distinguisher with a non-negligible probability of success.

6   A Family of Distinguishers

In Sect. 3 we found that the outputs at rounds 1, 3, 7 and 8 are biased allowing us to build a distinguisher. It is found that there exist plenty of 4-tuples of biased outputs. The generalization is presented in the following theorem.

Theorem 2. *The distribution of the outputs $(O_{r(i)}, O_{r+2(i+7)}, O_{t(i+7)}, O_{u(i+7)})$ of the TPy6 are biased for many suitably chosen (r, t, u)'s where $r > 0$; $t, u \geq 5$; $t \notin \{r, r + 2, u\}$; $u \notin \{r, r + 2, t\}$.*

We omit the proof as it is similar to the proof furnished for Theorem 1. This allows us to construct a family of distinguishers for the cipher TPy6. It seems possible to combine these huge number of distinguishers in order to construct one single efficient distinguisher; however, any concrete mathematical model to combine them is still an interesting open problem. Another major implication of the above generalization theorem is the fact that the TPy6 outputs will remain always biased no matter how many initial outputwords are discarded from the keystream.

7   Two New Ciphers: The TPy6-A and the TPy6-B

The Py-family of stream ciphers has been subject to extensive analysis ever since the Py and the Py6 were proposed in April 2005. The impressive speeds of the ciphers in software, particularly of the Py6 and the TPy6, have motivated us to modify the TPy6 to rule out all the attacks described in the previous sections of the paper. Firstly, many attacks on the Py, in particular, [4], [6], [7] and [14] can be easily translated to attacks on the Py6. However, due to smaller internal state of the TPy6, the attack described in [9] does not apply to the TPy6. Secondly, the speed of execution of Py6 on Pentium-III is about 2.82 cycles/byte which is very fast. These observations make the TPy6 and the Py6 more favorable to be used as fast stream ciphers than the Py. The TPy6 is resistant to the attacks described in [6], [14]. In order to generate a fast and secure stream cipher, we redesign the TPy6 where the variable rotation of a 32-bit term s in the round function is replaced by a constant, non-zero rotation term. The resultant cipher is named the TPy6-A. It is shown that this tweak clearly reduces one addition operation in each round (thereby, the performance is improved) and makes the cipher secure against all the existing attacks on the Py6 and the TPy6. A relatively slower version, where one outputword is removed from each round of the TPy6, is also

---

Algorithm 2. Round functions of the TPy6-A and the TPy6-B

Require: $Y[-3, ..., 64]$, $P[0, ..., 63]$, a 32-bit variable s
Ensure: 64-bit random (TPy6-A) or 32-bit random (TPy6-B) output
/*Update and rotate P*/
1. swap $(P[0], P[Y[43]\&63])$;
2. rotate (P);
   /* Update s*/
3. $s+ = Y[P[18]] - Y[P[57]]$;
4. $s = ROTL32(s, 19)$; /***Tweak:** variable rotation in the TPy6 replaced by a *constant non-zero* rotation*/
   /* Output 8 bytes (least significant byte first)*/
5. output $((ROTL32(s, 25) \oplus Y[64]) + Y[P[8]])$;/*this step is removed for TPy6-B*/
6. output $((\quad s \quad \oplus Y[-1]) + Y[P[21]])$;
   /* Update and rotate Y*/
7. $Y[-3] = (ROTL32(s, 14) \oplus Y[-3]) + Y[P[48]]$;
8. rotate(Y);

---

proposed. The speeds of execution of the TPy6-A and the TPy6-B on Pentium-III are 2.65 cycles/byte and 4.4 cycles/byte. Our security analysis conjectures that the TPy6-A and TPy6-B are immune to all attacks better than brute force. Algorithm 2 describes the PRBGs of the TPy6-A and the TPy6-B. Note that the key/IV setup algorithms of the ciphers are identical with the key/IV setup of the TPy6.

7.1 Security Analysis

Due to limit on space, we omit the security analysis of the ciphers, for which, the reader is kindly referred to the full version of this paper [12].

8 Conclusions and Open Problems

The first contribution of the paper is the development of a family of distinguishers from the outputs at rounds r, $r+2$, t and u of the cipher TPy6 (and Py6), where $r > 0$; $t, u \geq 5$; $t \notin \{r, r+2, u\}$; $u \notin \{r, r+2, t\}$. The best distinguisher works with data complexity $2^{224.6}$. It is reasonable to assume that these weak states can be combined to mount a more efficient attack on TPy; however, methods to combine many distinguishers into a single yet more efficient one is still an open problem. TPy6 and Py6. Exploiting the weaknesses in the key setup algorithm of the TPy6, a distinguisher with data complexity $2^{172.8}$ is built on the cipher. The second contribution is a proposal of two new, extremely fast stream ciphers TPy6-A and TPy6-B, which rule out all the existing attacks on the TPy6 and are conjectured to be immune to all attacks better than brute force.

References

1. Biham, E., Seberry, J.: Tweaking the IV Setup of the Py Family of Ciphers – The Ciphers Tpy, TPypy, and TPy6, January 25 (2007),
   http://www.cs.technion.ac.il/biham/
2. Biham, E., Seberry, J.: Py (Roo): A Fast and Secure Stream Cipher using Rolling Arrays. ecrypt submission (2005)
3. Biham, E., Seberry, J.: Pypy (Roopy): Another Version of Py. ecrypt submission (2006)
4. Crowley, P.: Improved Cryptanalysis of Py. In: Workshop Record of SASC 2006 - Stream Ciphers Revisited, ECRYPT Network of Excellence in Cryptology, Leuven, Belgium, pp. 52–60 (February 2006)
5. Mantin, I., Shamir, A.: A Practical Attack on Broadcast RC4. In: Matsui, M. (ed.) FSE 2001. LNCS, vol. 2355, pp. 152–164. Springer, Heidelberg (2002)
6. Isobe, T., Ohigashi, T., Kuwakado, H., Morii, M.: How to Break Py and Pypy by a Chosen-IV Attack. eSTREAM, ECRYPT Stream Cipher Project, Report (2006)/060
7. Paul, S., Preneel, B., Sekar, G.: Distinguishing Attacks on the Stream Cipher Py. In: Robshaw, M. (ed.) FSE 2006. LNCS, vol. 4047, pp. 405–421. Springer, Heidelberg (2006)

8. Paul, S., Preneel, B.: On the (In)security of Stream Ciphers Based on Arrays and Modular Addition. In: Lai, X., Chen, K. (eds.) ASIACRYPT 2006. LNCS, vol. 4284, pp. 69–83. Springer, Heidelberg (2006)
9. Sekar, G., Paul, S., Preneel, B.: Weaknesses in the Pseudorandom Bit Generation Algorithms of the Stream Ciphers TPypy and TPy. Cryptology ePrint Archive, Report 2007/ 075 (2007), http://eprint.iacr.org/2007/075.pdf
10. Sekar, G., Paul, S., Preneel, B.: New Weaknesses in the Keystream Generation Algorithms of the Stream Ciphers TPy and Py. In: Garay, J.A., Lenstra, A.K., Mambo, M., Peralta, R. (eds.) ISC 2007. LNCS, vol. 4779, pp. 249–262. Springer, Heidelberg (2007)
11. Sekar, G., Paul, S., Preneel, B.: Related-key Attacks on the Py-family of Ciphers and an Approach to Repair the Weaknesses. In: Srinathan, K., Rangan, C.P., Yung, M. (eds.) INDOCRYPT 2007. LNCS, vol. 4859, pp. 58–72. Springer, Heidelberg (2007)
12. Sekar, G., Paul, S., Preneel, B.: New Attacks on the Stream Cipher TPy6 and Design of New Ciphers the TPy6-A and the TPy6-B, Cryptology ePrint Archive, Report 2007/436, http://eprint.iacr.org/2007/436.pdf
13. Tsunoo, Y., Saito, T., Kawabata, T., Nakashima, H.: Distinguishing Attack against TPypy. In: Adams, C., Miri, A., Wiener, M. (eds.) SAC 2007. LNCS, vol. 4876. Springer, Heidelberg (2007),
http://dblp.uni-trier.de/rec/bibtex/conf/sacrypt/2007
14. Wu, H., Preneel, B.: Differential Cryptanalysis of the Stream Ciphers Py, Py6 and Pypy. In: Naor, M. (ed.) EUROCRYPT 2007. LNCS, vol. 4515, pp. 276–290. Springer, Heidelberg (2007)

Cryptanalysis of Achterbahn-128/80 with a New Keystream Limitation

María Naya-Plasencia*

Projet CODES, INRIA Paris-Rocquencourt, France
Maria.Naya_Plasencia@inria.fr

Abstract. This paper presents two key-recovery attacks against the modification of Achterbahn-128/80 proposed by the authors at SASC 2007 due to the previous attacks. The 80-bit variant, Achterbahn-80, was limited to produce at most 2^{52} bits of keystream with the same pair of key and IV, while Achterbahn-128 was limited to 2^{56} bits. The attack against Achterbahn-80 has complexity $2^{64.85}$ and needs fewer than 2^{52} bits of keystream, and the one against Achterbahn-128 has complexity 2^{104} and needs fewer than 2^{56} keystream bits. These attacks are based on the previous ones. The attack against Achterbahn-80 uses a new idea which allows us to reduce the required keystream length.

Keywords: eSTREAM, stream cipher, Achterbahn, cryptanalysis, correlation attack, linear approximation, parity check, key-recovery attack.

1 Introduction

The invention of public-key cryptography in the mid 1970's was a great progress. However, symmetric ciphers are still widely used because they are the only ones that can achieve high-speed or low-cost encryption. Today, we find symmetric ciphers in GSM mobile phones, in credit cards... Stream ciphers then form a subgroup of symmetric ciphers. In synchronous additive stream ciphers, the ciphertext is obtained by combining with a bitwise XOR the message with a secret binary sequence of the same length. This secret sequence is usually a pseudo-random one, that is generated with the help of a secret key by a pseudo-random generator, and it is called the keystream. Such pseudo-random generators are initialized by the secret key and they build in a deterministic way a long sequence that we cannot distinguish from a random one if we do not know the secret key. The eSTREAM project is a project launched by the European network ECRYPT about the conception of new stream ciphers. About thirty algorithms have been proposed in April 2005. Actually, in phase 3 of the project, 16 are

---

* This work was supported in part by the European Commission through the IST Programme under Contract IST-2002-507932 ECRYPT and by the ANR-06-SETI-013 project RAPIDE. The information in this document reflects only the author's views, is provided as is and no warranty is given that the information is fit for any particular purpose. The user thereof uses the information at its sole risk and liability.

S. Lucks, A.-R. Sadeghi, and C. Wolf (Eds.): WEWoRC 2007, LNCS 4945, pp. 142–152, 2008.

still being evaluated. Achterbahn [3, 5] is a stream cipher proposal submitted to the eSTREAM project that passed to phase 2 but not to phase 3 of eS-TREAM. After the cryptanalysis of the first two versions [8, 10], it moved on to a new one called Achterbahn-128/80 [4] published in June 2006. Achterbahn-128/80 corresponds to two keystream generators with key sizes of 128 bits and 80 bits, respectively. Their maximal keystream length was limited to 2^{63}, but, in order to avoid the attacks presented in [9, 11], the maximal keystream length was re-limited to produce at most 2^{52} bits of keystream with the same pair of key and IV for Achterbahn-80, while Achterbahn-128 was limited to 2^{56} bits. This paper presents two key-recovery attacks against this modification to Achterbahn-128/80, proposed by the authors at SASC 2007 [6]. The attack against Achterbahn-80 has complexity $2^{64.85}$ and needs fewer than 2^{52} bits of keystream, and the one against Achterbahn-128 has complexity 2^{104} and needs fewer than 2^{56} keystream bits. These attacks are based on the previous ones. The attack against Achterbahn-80 uses a new idea which allows us to reduce the required keystream length.

The paper is organized as follows. Section 2 presents the main specifications of Achterbahn-128/80. Section 3 then describes a distinguishing attack against Achterbahn-80. Section 4 presents a distinguishing attack against Achterbahn-128. Section 5 describes how this previous distinguishing attacks can be transformed into key-recovery attacks.

2   Achterbahn-128/80

Achterbahn-128 and Achterbahn-80 are composed of a number of feedback shift registers whose outputs are taken as inputs of a Boolean combining function and where the keystream is the output of this function at each instant t.

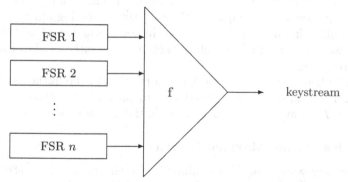

2.1   Main Specifications of Achterbahn-128

Achterbahn-128 consistis of 13 binary nonlinear feedback shift registers (NLF-SRs) denoted by $R0, R1, \ldots, R12$. The length of register i is $L_i = 21 + i$ for $i = 0, 1, \ldots, 12$. These NLFSRs are primitive in the sense that their periods T_i are equal to $2^{L_i} - 1$. Each sequence which is used as an input to the Boolean combining function is not the output sequence of the NLFSR directly, but a shifted version of itself. The shift amount depends on the register number, but it is fixed for each register. In the following, $x_i = (x_i(t))_{t \geq 0}$ for $0 \leq i \leq 12$ denotes the shifted version of the output of the register i at time t. The output of the keystream generator at time t, denoted by $S(t)$, is the one of the Boolean combining function F with the inputs corresponding to the output sequences of the NLFSRs correctly shifted, i.e. $S(t) = F(x_0(t), \ldots, x_{12}(t))$. The algebraic normal form of the 13-variable combining function F is given in [4]. Its main cryptographic properties are: balancedness, algebraic degree 4, correlation immunity order 8, nonlinearity 3584, algebraic immunity 4.

2.2 Main Specifications of Achterbahn-80

Achterbahn-80 consists of 11 registers, which are the same ones as in the above case, except for the first and the last ones. The Boolean combining function, G, is a sub-function of F :

$$G(x_1, \ldots, x_{11}) = F(0, x_1, \ldots, x_{11}, 0).$$

Its main cryptographic properties are: balancedness, algebraic degree 4, correlation immunity order 6, nonlinearity 896, algebraic immunity 4. As we can see, Achterbahn-128 contains Achterbahn-80 as a substructure.

2.3 The Key-Loading Algorithm

The key-loading algorithm uses the key K of 128/80 bits and an initial value IV of 128/80 bits. The method for initializing the registers is the following one: first of all, all registers are filled with the bits of $K||IV$. After that, Register i is clocked $a - L_i$ times where a is the number of bits of $K||IV$, and the remaining bits of $K||IV$ are added to the feedback bit. Then, each register outputs one bit. Those bits are taken as inputs of the Boolean combining function, which outputs a new bit. This bit is now added to the feedbacks for 32 additional clockings. Then we overwrite the last cell of each register with a 1, in order to avoid the all zero state.

This algorithm has been modified in relation to the initial versions of Achterbahn. The aim of this modification is to prevent the attacker from recovering the key K from the knowledge of the initial states of some registers.

2.4 Keystream Maximal Length

In the first version of Achterbahn-128/80, the maximal keystream length was limited to 2^{63}. As this version was attacked [9, 11], the authors proposed a new limitation of the keystream length [6], which was 2^{52} for Achterbahn-80 and 2^{56} for Achterbahn-128. We present here two attacks against both generators, which are based on the previous ones. The attack against the 80-bit variant, Achterbahn-80, has complexity $2^{64.85}$ and needs fewer than 2^{52} keystream bits. The attack against Achterbahn-128 requires 2^{104} operations and fewer than 2^{56} keystream bits.

3 Distinguishing Attack against Achterbahn-80

Now, we describe a new attack against Achterbahn-80 with a complexity of $2^{64.85}$ where a linear approximation of the output function is considered. The attack is a distinguishing attack but it also allows to recover the initial states of certain constituent registers.

This attack is very similar to the previous attack against Achterbahn-80 presented in [11]. It relies on a biased parity-check relation between the keystream bits which holds with probability

$$p = \frac{1}{2}(1 + \eta) \text{ with } |\eta| \ll 1,$$

where η is the bias of the relation. The attack exploits an s-variable linear approximation ℓ of the combining function G. For now on we denote by $T_{i,j}$ the least common multiple of the periods of Registers i and j. We build the parity-check equations, as the ones introduced in [10] and used in [11] derived from ℓ:

$$\ell(t) = \sum_{j=1}^{s} x_{i_j}(t)$$

at 2^m different instants $(t+\tau)$, where τ varies in the set of the linear combinations with $0-1$ coefficients of $T_{i_1,i_2}, T_{i_3,i_4}, \ldots, T_{i_{2m-1},i_{2m}}$. In the following, this set is denoted by $\langle T_{i_1,i_2}, \ldots, T_{i_{2m-1},i_{2m}} \rangle$, i.e.,

$$\mathcal{I} = \langle T_{i_1,i_2}, \ldots, T_{i_{2m-1},i_{2m}} \rangle = \left\{ \sum_{j=1}^{m} c_j T_{i_{2j-1},i_{2j}}, c_1, \ldots, c_m \in \{0,1\} \right\}.$$

We know that:

$$\sum_{\tau \in \mathcal{I}} x_{i_1}(t+\tau) + \ldots + x_{i_{2m}}(t+\tau) = 0,$$

this leads to a parity-check sequence $\ell\ell$ defined by:

$$\ell\ell(t) = \sum_{\tau \in \mathcal{I}} \ell(t+\tau) = \sum_{\tau \in \mathcal{I}} \left(x_{i_{2m+1}}(t+\tau) + \ldots + x_{i_s}(t+\tau) \right).$$

Note that each term with index i_{2j-1} is associated to the corresponding term i_{2j} to build the parity check, because it enables us to eliminate the influence of $2m$ registers in a parity-check with 2^m terms only.

Approximation of the combining function. Following this general principle, our attack exploits the following linear approximation of the combining function G:

$$\ell(x_1, \ldots, x_{11}) = x_1 + x_3 + x_4 + x_5 + x_6 + x_7 + x_{10}.$$

It is worth noticing that, since the combining function G is 6-resilient, any approximation of G involves at least 7 input variables. Moreover, the highest bias corresponding to an approximation of G by a 7-variable function is achieved by a function of degree one as proved in [2].

For $\ell(t) = x_1(t) + x_3(t) + x_4(t) + x_5(t) + x_6(t) + x_7(t) + x_{10}(t)$, the keystream $(S(t))_{t \geq 0}$ satisfies $\Pr[S(t) = \ell(t)] = \frac{1}{2}(1 - 2^{-3})$.

Parity-checks. Let us build a parity-check as follows:

$$\ell\ell(t) = \ell(t) + \ell(t + T_{3,7}) + \ell(t + T_{4,5}) + \ell(t + T_{3,7} + T_{4,5}).$$

Therefore, this corresponds to $s = 7$ and $m = 2$ in the general description of the attack. The terms containing the sequences x_3, x_4, x_5, x_7 vanish in $\ell\ell(t)$, so $\ell\ell(t)$ depends exclusively on the sequences x_1, x_6 and x_{10}. Thus, we have

$$\ell\ell(t) = \sigma_1(t) + \sigma_6(t) + \sigma_{10}(t),$$

where

$$\sigma_i(t) = x_i(t) + x_i(t + T_{3,7}) + x_i(t + T_{4,5}) + x_i(t + T_{3,7} + T_{4,5}).$$

The period $T_{4,5}$ is 2^{51} and the period $T_{3,7}$ is smaller than 2^{49} as T_3 and T_7 have common factors, so to build those parity checks we need less than the maximal keystream length allowed.

Adding four times the approximation has the effect of multiplying the bias four times, so

$$\sigma(t) = S(t) + S(t + T_{3,7}) + S(t + T_{4,5}) + S(t + T_{3,7} + T_{4,5})$$

where $(S(t))_{t \geq 0}$ is the keystream satisfies

$$Pr[\sigma(t) = \sigma_1(t) + \sigma_6(t) + \sigma_{10}(t)] = \frac{1}{2}(1 + \eta)$$

with $\eta = 2^{-4 \times 3}$.

We now decimate $\sigma(t)$ by the period of Register 1, which is involved in the parity-check, so we create like this a new parity-check:

$$\sigma'(t) = \sigma(t(2^{22} - 1)).$$

Now, we have that $\sigma'(t)$ is an approximation of $(\sigma_6(t(2^{22} - 1)) + \sigma_{10}(t(2^{22} - 1)))$ with biais $+\eta$ or $-\eta$. Then, if we did as in the previous attack in [11], the one before the new keystream limitation, where we performed an exhaustive search for the initial states of Registers 6 and 10, we would need

$$2^{3 \times 4 \times 2} \times 2 \times (58 - 2) \times \ln(2) = 2^{30.29}$$

parity-checks $\sigma'(t)$ to detect this bias. As we are decimating by the period of the Register 1, we would need $2^{30.29} \times 2^{22} = 2^{52.29}$ keystream bits to perform the attack, and it is over the limitation, so we cannot do that.

In the previous attack we took only the first bit of the keystream and decimated by the period of the first register $2^{30.29}$ times. What we do now is to consider the first four consecutive shifts of the keystream and for each one, we obtain a sequence of $\frac{2^{30.29}}{4} = 2^{28.29}$ bits by decimating it by the period of the first register $2^{28.29}$ times. Thus, we consider the first $2^{50.29}$ bits of the keystream and we compute the $4 \times 2^{28.29} = 2^{30.29}$ parity checks:

$$S\left(t(2^{22} - 1) + i\right) + S(t(2^{22} - 1) + i + T_{3,7}) + S(t(2^{22} - 1) + i + T_{4,5}) +$$
$$S\left(t(2^{22} - 1) + i + T_{3,7} + T_{4,5}\right)$$

for $i \in \{0, \ldots, 3\}$ and $0 \le t < 2^{28.29}$. This way, the required number of keystream bits is reduced to $2^{28.29} \times 2^{22} = 2^{50.29}$ and respects the maximal keystream length permitted.

Thus, we perform an exhaustive search over Registers 6 and 10, adapting to our new situation the algorithm introduced in [11]. We will have to compute, for each one of the previously mentioned sequences, so for each $i \in \{0, 1, 2, 3\}$, the following sum:

$$S = \sum_{t'=0}^{2^{28.29}-1} \sigma(t'T_1 + i) \oplus \ell\ell(t'T_1 + i)$$

Using the decomposition

$$2^{28.29} = 2T_6 + T' \text{ with } T' = 2^{25.83},$$

we obtain

$$
\begin{aligned}
S &= \sum_{t'=0}^{2^{28.29}-1} \sigma(t'T_1 + i) \oplus \ell\ell(t'T_1 + i) \\
&= \sum_{k=0}^{T'} \sum_{t=0}^{2} \sigma((T_6 t + k)T_1 + i) \oplus \ell\ell((T_6 t + k)T_1 + i) \\
&+ \sum_{k=T'+1}^{T_6-1} \sum_{t=0}^{1} \sigma((T_6 t + k)T_1 + i) \oplus \ell\ell((T_6 t + k)T_1 + i) \\
&= \sum_{k=0}^{T'} \sum_{t=0}^{2} \sigma((T_6 t + k)T_1 + i) \oplus \sigma_{10}((T_6 t + k)T_1 + i) \oplus \sigma_6((T_6 t + k)T_1 + i) \\
&+ \sum_{k=T'+1}^{T_6-1} \sum_{t=0}^{1} \sigma((T_6 t + k)T_1 + i) \oplus \sigma_{10}((T_6 t + k)T_1 + i) \oplus \sigma_6((T_6 t + k)T_1 + i)
\end{aligned}
$$

$$
\begin{aligned}
&= \sum_{k=0}^{T'} \left[(\sigma_6(kT_1 + i) \oplus 1) \left(\sum_{t=0}^{2} \sigma((T_6 t + k)T_1 + i) \oplus \sigma_{10}((T_6 t + k)T_1 + i) \right) \right. \\
&\left. + \sigma_6(kT_1 + i) \left(3 - \sum_{t=0}^{2} \sigma((T_6 t + k)T_1 + i) \oplus \sigma_{10}((T_6 t + k)T_1 + i) \right) \right] \\
&+ \sum_{k=T'}^{T_6-1} \left[(\sigma_6(kT_1 + i) \oplus 1) \left(\sum_{t=0}^{1} \sigma((T_6 t + k)T_1 + i) \oplus \sigma_{10}((T_6 t + k)T_1 + i) \right) \right. \\
&\left. + \sigma_6(kT_1 + i) \left(2 - \sum_{t=0}^{1} \sigma((T_6 t + k)T_1 + i) \oplus \sigma_{10}((T_6 t + k)T_1 + i) \right) \right],
\end{aligned}
$$

where $\sigma(t), \sigma_6(t)$ and $\sigma_{10}(t)$ are the parity checks computed at the instant t with the keystream, the sequence generated by Register 6 and the one generated by

Register 10 respectively. Note that we have to split the sum at T', because for $k \leq T'$ we have to sum the parity checks at 3 instants but for $k > T'$ we only have to sum them at 2 instants, since $T' + 2 \times T_6 = 2^{28.29}$. The sum can be written in the previous way since $\sigma_6((T_6t + k)T_1 + i)$ is constant for fixed values of k and i. The attack then consists of the following steps:

- We choose an initial state for Register 6, e.g. the all-one initial state. We compute and save a binary vector V_6 of length T_6, $V_6[k] = \sigma_6(k)$, where the sequence with whom we are computing $\sigma_6(k)$ is generated from the chosen initial state. The complexity of this state is $T_6 \times 2^2$ operations.
- For each possible initial state of Register 10 (so 2^{31-1} possibilities):
  - we compute and save four vectors $V_{10,i}$, where $i \in \{0, 1, 2, 3\}$, each one composed of T_6 integers of 2 bits.

$$V_{10,i}[k] = \sum_{t=0}^{q} \sigma((T_6t + k)T_1 + i) \oplus \sigma_{10}((T_6t + k)T_1 + i),$$

  where $q = 2$ if $k \leq T'$ and $q = 1$ if $k > T'$. The time complexity of this step is:

$$2^2 \left(3 \times 2^{25.83} + 2(2^{27} - 1 - 2^{25.83})\right)(7 + 2) = 2^2 \times 2^{28.29} \times 2^{3.1} = 2^{33.49}$$

  for each possible initial state of Register 10, where 2^2 is the number of vectors that we are computing, 7 corresponds to the number of operations required for computing each $(\sigma(t) + \sigma_{10}(t))$ and $2^{28.29} \times 2$ is the cost of summing up $2^{28.29}$ integers of 2 bits.
  - For each possible p from 0 to $T_6 - 1$:
    * we define $V_{6,i}$ of length T_6, $\forall i \in \{0, 1, 2, 3\}$: $V_{6,i}[k] = V_6[k + p + i \mod T_6]$.
      Actually, $(V_{6,i}[k])_{k<T_6}$ corresponds to $(\sigma_6(k))_{k<T_6}$ when the initial state of Register 6 corresponds to the internal state obtained after clocking $(i + p)$ times Register 6 from the all-one initial state.
    * With the eight vectors that we have obtained

$$(V_{10,0}, \ldots, V_{10,3}, V_{6,0}, \ldots, V_{6,3}),$$

    we compute for each $i \in \{0, 1, 2, 3\}$:

$$W_i = \sum_{k=0}^{T'} [(V_{6,i}[k] \oplus 1) V_{10,i}[k] + V_{6,i}[k] (3 - V_{10,i}[k])] +$$
$$\sum_{k=T'+1}^{T_6-1} [(V_{6,i}[k] \oplus 1) V_{10,i}[k] + V_{6,i}[k] (2 - V_{10,i}[k])].$$

When we do this with the correct initial states of Registers 6 and 10, we will find an important bias for the four W_i.

The complexity of this point would be, for each p $2^2 \times T_6 \times 8 = 2^{32}$, so $2^{32} \times 2^{27} = 2^{59}$. The total number of memory accesses for each possible initial state of Register 10 is

$$4 \times T_6 + 4 \times 2 \times T_6 = 2^{30.3},$$

where the first term corresponds to the storage of $V_{10,i}$ and the second one corresponds to the accesses to $V_{6,i}$ and $V_{10,i}$ to compute W_i. But we can speed up the process by defining a new vector,

$$V'_{10,j}[k] = V_{10,j}[k] + ct$$

where $ct = 0$ if $k \leq T'$ and $ct = 0.5$ if $k > T'$.
Then, for each i we are going to compute:

$$\sum_{k'=0}^{T_6-1} (-1)^{V_{6,i}[k+p]} \left(V'_{10,i}[k] - \frac{3}{2} \right) + (T' \times 1.5 + (T_6 - T') \times 1).$$

The issue is now to find the p that maximizes this sum, this is the same as computing the maximum of the crosscorrelation of two sequences of length T_6. We can do that efficiently using a fast Fourier transform as explained in [1, pages 306-312]. The final complexity for computing this sum will be in $T_6 \log_2(T_6)$.

Thus, the total complexity of this state will be $4T_6 \log_2(T_6) \approx 2^{34}$.

We now compute the false alarm and the non detection probabilities. First of all we consider as the bias threshold $S = 0.55 \times \eta = 2^{-12.86}$. Let n be the length of the sequences used and i be the number of sequences. The false alarm probability for i sequences is the probability that, while trying wrong initial states of Registers 6 and 10 (which would generate random sequences) we find a bias higher than $2^{-12.86}$ or lower than $-2^{-12.86}$ for all the i W_j. Using Chernoff's bound on the tail of the binomial distribution we get:

$$P_{fa4}(S) = (P_{fa1}(S))^i \leq (2e^{-2S^2n})^i,$$

where P_{fa1} is the false alarm probability for one sequence. In our case $n = 2^{28.29}$ and $i = 4$, so $P_{fa4} = 2^{-64.29}$. The number of initial states that will pass the test without being the correct one will be $(2^{56} - 1) \times 2^{-64.29} = 2^{-8.29}$. The non detection probability is the probability that while trying the correct initial states of Registers 6 and 10 we find a bias between $-2^{-12.86}$ and $2^{-12.86}$. For one sequence it will be:

$$P_{nd1}(S) \leq 2e^{-2(\eta-S)^2n},$$

So the probability of non detection for i sequences, that is, the probability of not detecting the threshold at one of the i W_j will be:

$$P_{nd4}(S) = 1 - (1 - P_{nd1}(S))^i.$$

It is used only for the correct initial states. As we can see, it increases with i, the number of used sequences. In our case, $i = 4$, leading to $P_{nd4}(S) = 2^{-9.43}$.

The time complexity is, finally

$$2^{L_{10}-1} \times \left[2^{33.69} + 4T_6 \log_2 4T_6\right] + T_6 \times 2^2 = 2^{64.85} \text{ steps.}$$

The required number of keystream bits is

$$2^{28.29} \times T_1 + T_3 T_7 + T_4 T_5 = 2^{50.29} + 2^{48.1} + 2^{51} < 2^{52}.$$

The memory used is

$$2^{30} + 2^{29} = 2^{30.58},$$

where 2^{30} is the size of the four $V_{10,i}$ vectors and 2^{29} of the $V_{6,i}$ vectors.

4   Distinguishing Attack against Achterbahn-128

Now, we present a distinguishing attack against the 128-bit version of Achterbahn which also recovers the initial states of two registers.

We consider the following approximation of the combining function F:

$$\ell(x_0, \dots, x_{12}) = x_0 + x_1 + x_2 + x_3 + x_4 + x_7 + x_8 + x_9 + x_{10}.$$

Then, for $\ell(t) = x_0(t) + x_1(t) + x_2(t) + x_3(t) + x_4(t) + x_7(t) + x_8(t) + x_9(t) + x_{10}(t)$, we have $\Pr[S(t) = \ell(t)] = \frac{1}{2}(1 + 2^{-3})$.

Parity-checks. If we build a parity check as follows:

$$\ell\ell\ell(t) = \sum_{\tau \in \langle T_{3,8}, T_{1,10}, T_{2,9} \rangle} \ell(t + \tau),$$

the terms containing the sequences x_1, x_2, x_3, x_8, x_9, x_{10} will disappear from $\ell\ell\ell(t)$, so $\ell\ell\ell(t)$ depends exclusively on the sequences x_0, x_4 and x_7:

$$\ell\ell\ell(t) = \sum_{\tau \in \langle T_{3,8}, T_{1,10}, T_{2,9} \rangle} x_0(t + \tau) + x_4(t + \tau) + x_7(t + \tau) = \sigma_0(t) + \sigma_4(t) + \sigma_7(t),$$

where $\sigma_0(t)$, $\sigma_4(t)$ and $\sigma_7(t)$ are the parity-checks computed on the sequences generated by Registers 0, 4 and 7. Adding eight times the approximation has the effect of multiplying the bias eight times, so the bias of

$$\sigma(t) = \sum_{\tau \in \langle T_{3,8}, T_{1,10}, T_{2,9} \rangle} S(t + \tau),$$

where $(S(t))_{t\geq 0}$ is the keystream, is $2^{-8\times 3}$. So:

$$\Pr[\sigma(t) + \sigma_0(t) + \sigma_4(t) + \sigma_7(t) = 1] = \frac{1}{2}(1 - \varepsilon^8).$$

This means that we need $2^{3\times 8\times 2} \times 2 \times (74 - 3) \times \ln(2) = 2^{54.63}$ values of $\sigma(t) + \sigma_0(t) + \sigma_4(t) + \sigma_7(t)$ to detect this bias, when we perform an exhaustive search on Registers 0, 4 and 7.

We use the previously proposed algorithm for the attack of Achterbahn-128 for computing the sum $\sigma(t) + \sigma_0(t) + \sigma_4(t) + \sigma_7(t)$ over all values of t. This algorithm has a lower complexity than an exhaustive search for the initial states of the Registers 0, 4 and 7 simultaneously. We use it considering Register 0 and Register 4 together.

The complexity is going to be, finally

$$2^{L_0-1} \times 2^{L_4-1} \times \left[2^{54.63} \times \left(2^4 + 2^{4.7}\right) + T_7 \log T_7\right] + T_7 \times 2^3 = 2^{104} \text{ steps}.$$

The required keystream length is:

$$2^{54.63} + T_{1,10} + T_{2,9} + T_{3,8} = 2^{54.63} + 2^{53} + 2^{53} + 2^{53} < 2^{56} \text{ bits}.$$

The memory used is

$$2^{32} + 2^{28} = 2^{32.08},$$

where 2^{32} is the size of the V_{0-4} vector and 2^{28} of the V_7 vector.

5   Recovering the Key

As explained in the previous attacks [11] and introduced in [9], we can recover the key with a variant of a meet-in-the-middle attack once we have found the initial state of some registers. The time complexity of this part of the attack is smaller than the one of the previously described distinguishing attack that we need to get the initial states of several registers. So the complexity of the total key-recovery attack is the same one as for the distinguishing attacks.

6   Conclusion

We have proposed an attack against Achterbahn-80 in $2^{64.85}$ steps where fewer than 2^{52} bits are needed. That is $2^{64.85}$ boolean operations, which makes it much more efficient than a brute force attack. The memory needed for this attack is $2^{30.58}$. An attack against Achterbahn-128 is also proposed in 2^{104} steps where fewer than 2^{56} bits of keystream are required. The memory needed is 2^{32}. After that we can recover the key of Achterbahn-80 with a complexity of 2^{40} in time and 2^{41} in memory (the time complexity is less than for the distinguishing part of the attack). For Achterbahn-128 we can recover the key with a complexity of 2^{73} in time and 2^{48} in memory. After those attacks, the authors proposed a new keystream limitation for both Achterbahn-128/80 [7]. This new limitation is 2^{44}. With this limitation the known attacks are not applicable.

References

1. Blahut, R.E.: Fast Algorithms for Digital Signal Processing. Addison Wesley, Reading (1985)
2. Canteaut, A., Trabbia, M.: Improved Fast Correlation Attacks Using Parity-Check Equations of Weight 4 and 5. In: Preneel, B. (ed.) EUROCRYPT 2000. LNCS, vol. 1807, pp. 573–588. Springer, Heidelberg (2000)
3. Gammel, B.M., Gottfert, R., Kniffler, O.: The Achterbahn stream cipher. eSTREAM, ECRYPT Stream Cipher Project, Report 2005/002 (2005), http://www.ecrypt.eu.org/stream/ciphers/achterbahn/achterbahn.pdf
4. Gammel, B.M., Gottfert, R., Kniffler, O.: Achterbahn-128/80. eSTREAM, ECRYPT Stream Cipher Project, Report 2006/001 (2006), http://www.ecrypt.eu.org/stream/p2ciphers/achterbahn/achterbahn_p2.pdf
5. Gammel, B.M., Gottfert, R., Kniffler, O.: Status of Achterbahn and tweaks. eSTREAM, ECRYPT Stream Cipher Project, Report 2006/027 (2006), http://www.ecrypt.eu.org/stream/papersdir/2006/027.pdf
6. Gammel, B.M., Gottfert, R., Kniffler, O.: Achterbahn-128/80: Design and analysis. In: ECRYPT Network of Excellence - SASC Workshop Record, pp. 152–165 (2007)
7. Gammel, B.M., Gottfert, R.: On the frame length of Achterbahn-128/80. In: IEEE Information Theory Workshop on Information Theory for Wireless Networks, pp. 91–95 (2007)
8. Hell, M., Johansson, T.: Cryptanalysis of Achterbahn-version 2. In: Biham, E., Youssef, A.M. (eds.) SAC 2006. LNCS, vol. 4356, pp. 45–55. Springer, Heidelberg (2007)
9. Hell, M., Johansson, T.: Cryptanalysis of Achterbahn-128/80. IET Information Security 1(2) (2007)
10. Johansson, T., Meier, W., Muller, F.: Cryptanalysis of Achterbahn. In: Robshaw, M. (ed.) FSE 2006. LNCS, vol. 4047, pp. 1–14. Springer, Heidelberg (2006)
11. Naya-Plasencia, M.: Cryptanalysis of Achterbahn-128/80. In: Biryukov, A. (ed.) FSE 2007. LNCS, vol. 4593, pp. 73–86. Springer, Heidelberg (2007)

Author Index

Lecture Notes in Computer Science

Sublibrary 4: Security and Cryptology

For information about Vols. 1– 4176
please contact your bookseller or Springer

Vol. 4817: K.-H. Nam, G. Rhee (Eds.), Information Security and Cryptology - ICISC 2007. XIII, 367 pages. 2007.

Vol. 4812: P. McDaniel, S.K. Gupta (Eds.), Information Systems Security. XIII, 322 pages. 2007.

Vol. 4784: W. Susilo, J.K. Liu, Y. Mu (Eds.), Provable Security. X, 237 pages. 2007.

Vol. 4779: J.A. Garay, A.K. Lenstra, M. Mambo, R. Peralta (Eds.), Information Security. XIII, 437 pages. 2007.

Vol. 4776: N. Borisov, P. Golle (Eds.), Privacy Enhancing Technologies. X, 273 pages. 2007.

Vol. 4752: A. Miyaji, H. Kikuchi, K. Rannenberg (Eds.), Advances in Information and Computer Security. XIII, 460 pages. 2007.

Vol. 4734: J. Biskup, J. López (Eds.), Computer Security – ESORICS 2007. XIV, 628 pages. 2007.

Vol. 4727: P. Paillier, I. Verbauwhede (Eds.), Cryptographic Hardware and Embedded Systems - CHES 2007. XIV, 468 pages. 2007.

Vol. 4691: T. Dimitrakos, F. Martinelli, P.Y.A. Ryan, S. Schneider (Eds.), Formal Aspects in Security and Trust. VIII, 285 pages. 2007.

Vol. 4677: A. Aldini, R. Gorrieri (Eds.), Foundations of Security Analysis and Design IV. VII, 325 pages. 2007.

Vol. 4657: C. Lambrinoudakis, G. Pernul, A.M. Tjoa (Eds.), Trust, Privacy and Security in Digital Business. XIII, 291 pages. 2007.

Vol. 4637: C. Kruegel, R. Lippmann, A. Clark (Eds.), Recent Advances in Intrusion Detection. XII, 337 pages. 2007.

Vol. 4631: B. Christianson, B. Crispo, J.A. Malcolm, M. Roe (Eds.), Security Protocols. IX, 347 pages. 2007.

Vol. 4622: A. Menezes (Ed.), Advances in Cryptology - CRYPTO 2007. XIV, 631 pages. 2007.

Vol. 4593: A. Biryukov (Ed.), Fast Software Encryption. XI, 467 pages. 2007.

Vol. 4586: J. Pieprzyk, H. Ghodosi, E. Dawson (Eds.), Information Security and Privacy. XIV, 476 pages. 2007.

Vol. 4582: J. López, P. Samarati, J.L. Ferrer (Eds.), Public Key Infrastructure. XI, 375 pages. 2007.

Vol. 4579: B.M. Hämmerli, R. Sommer (Eds.), Detection of Intrusions and Malware, and Vulnerability Assessment. X, 251 pages. 2007.

Vol. 4575: T. Takagi, T. Okamoto, E. Okamoto, T. Okamoto (Eds.), Pairing-Based Cryptography – Pairing 2007. XI, 408 pages. 2007.

Vol. 4567: T. Furon, F. Cayre, G. Doërr, P. Bas (Eds.), Information Hiding. XI, 393 pages. 2008.

Vol. 4521: J. Katz, M. Yung (Eds.), Applied Cryptography and Network Security. XIII, 498 pages. 2007.

Vol. 4515: M. Naor (Ed.), Advances in Cryptology - EUROCRYPT 2007. XIII, 591 pages. 2007.

Vol. 4499: Y.Q. Shi (Ed.), Transactions on Data Hiding and Multimedia Security II. IX, 117 pages. 2007.

Vol. 4464: E. Dawson, D.S. Wong (Eds.), Information Security Practice and Experience. XIII, 361 pages. 2007.

Vol. 4462: D. Sauveron, K. Markantonakis, A. Bilas, J.-J. Quisquater (Eds.), Information Security Theory and Practices. XII, 255 pages. 2007.

Vol. 4450: T. Okamoto, X. Wang (Eds.), Public Key Cryptography – PKC 2007. XIII, 491 pages. 2007.

Vol. 4437: J.L. Camenisch, C.S. Collberg, N.F. Johnson, P. Sallee (Eds.), Information Hiding. VIII, 389 pages. 2007.

Vol. 4392: S.P. Vadhan (Ed.), Theory of Cryptography. XI, 595 pages. 2007.

Vol. 4377: M. Abe (Ed.), Topics in Cryptology – CT-RSA 2007. XI, 403 pages. 2006.

Vol. 4356: E. Biham, A.M. Youssef (Eds.), Selected Areas in Cryptography. XI, 395 pages. 2007.

Vol. 4341: P.Q. Nguyên (Ed.), Progress in Cryptology - VIETCRYPT 2006. XI, 385 pages. 2006.

Vol. 4332: A. Bagchi, V. Atluri (Eds.), Information Systems Security. XV, 382 pages. 2006.

Vol. 4329: R. Barua, T. Lange (Eds.), Progress in Cryptology - INDOCRYPT 2006. X, 454 pages. 2006.

Vol. 4318: H. Lipmaa, M. Yung, D. Lin (Eds.), Information Security and Cryptology. XI, 305 pages. 2006.

Vol. 4307: P. Ning, S. Qing, N. Li (Eds.), Information and Communications Security. XIV, 558 pages. 2006.

Vol. 4301: D. Pointcheval, Y. Mu, K. Chen (Eds.), Cryptology and Network Security. XIII, 381 pages. 2006.

Vol. 4300: Y.Q. Shi (Ed.), Transactions on Data Hiding and Multimedia Security I. IX, 139 pages. 2006.

Vol. 4298: J.K. Lee, O. Yi, M. Yung (Eds.), Information Security Applications. XIV, 406 pages. 2007.

Vol. 4296: M.S. Rhee, B. Lee (Eds.), Information Security and Cryptology – ICISC 2006. XIII, 358 pages. 2006.

Vol. 4284: X. Lai, K. Chen (Eds.), Advances in Cryptology – ASIACRYPT 2006. XIV, 468 pages. 2006.

Vol. 4283: Y.Q. Shi, B. Jeon (Eds.), Digital Watermarking. XII, 474 pages. 2006.

Vol. 4266: H. Yoshiura, K. Sakurai, K. Rannenberg, Y. Murayama, S.-i. Kawamura (Eds.), Advances in Information and Computer Security. XIII, 438 pages. 2006.

Vol. 4258: G. Danezis, P. Golle (Eds.), Privacy Enhancing Technologies. VIII, 431 pages. 2006.

Vol. 4249: L. Goubin, M. Matsui (Eds.), Cryptographic Hardware and Embedded Systems - CHES 2006. XII, 462 pages. 2006.

Vol. 4237: H. Leitold, E.P. Markatos (Eds.), Communications and Multimedia Security. XII, 253 pages. 2006.

Vol. 4236: L. Breveglieri, I. Koren, D. Naccache, J.-P. Seifert (Eds.), Fault Diagnosis and Tolerance in Cryptography. XIII, 253 pages. 2006.

Vol. 4219: D. Zamboni, C. Krügel (Eds.), Recent Advances in Intrusion Detection. XII, 331 pages. 2006.

Vol. 4189: D. Gollmann, J. Meier, A. Sabelfeld (Eds.), Computer Security – ESORICS 2006. XI, 548 pages. 2006.